AF601235

Aquatic Food Safety and Quality Management

THE AUTHORS

Mr F Parthiban is the Assistant Professor of Tamil Nadu Dr. J. Jayalalithaa Fisheries University, Nagapattinam. He has 10 years of fisheries industrial experience as managers in aquaculture, fish processing, and quality control at Saudi Arabia, Nigeria, and Tanzania and 4 years of experience in teaching, research and extension programmes. He has Recipient of CIFE, ICAR fellowship for Master of Fisheries Science in Post Harvest Technology at Central Institute of Fisheries Technology. He has successfully completed internal auditor and lead auditor courses for Food Safety Management System, ISO 22000:2005. He is the member of society of Fisheries Technologist and Professional Fisheries Graduates Forum in India. He has good expertise in aquatic food processing, aquatic food safety and quality. He has been handling courses for fisheries graduate on Fundamentals of Biochemistry , Biochemical Techniques and Instrumentation, Microbiology of fish and fishery product, Fish Processing Technology. He has been conferred with "Best Teacher Award" by the Tamil Nadu Dr. J. Jayalalithaa Fisheries University for the year 2017. He has published research and extension articles in national and international journal.

Dr S Balasundari is working at Tamil Nadu Dr. M.G.R. Fisheries College & Research Institute, Tamil Nadu Dr. J. Jayalalithaa Fisheries University at Ponneri as Professor and Head, Department of Fish Processing Technology, having 25 years of professional experience that includes her service in fish processing industries, ARS and SAU. She is a member of AFSIB, Agricultural Scientific Tamil Society, SOFT and WAS. She served as a production executive in Seafood Processing and Export Companies for three years, possessing sound knowledge on processing and export of shrimp, crab, lobster, squid, cuttlefish and finfish. She has developed various contemporary value added fish products and disseminated the technologies. She has completed six projects through state and national level funding agencies and established demonstration units for the benefit of stakeholders.

Dr B Ahilan, is working as Dean in charge at Tamil Nadu Dr. M.G.R. Fisheries College & Research Institute, Tamil Nadu Dr. J. Jayalalithaa Fisheries University, Ponneri. He did his Ph.D., in Aquaculture and has put up a total service of 28 years in the University. He is teaching Aquaculture related courses for Under graduate, post graduate and doctoral students of Fisheries science. He has guided forty eight post graduate students, both M.F.Sc. and Ph.D in the capacity of chairman / member. He has received Best Teacher Award sponsored by PFGF, Mumbai. He has operated numerous external funded research projects related to aquaculture. He has a total of 276 publications to his credit which includes 84 research articles, 144 popular articles, 9 text books, 14 manuals and 25 pamphlets. He has organized and co-ordinated several training programmes in fish culture for the benefit of fish farmers, rural folks and entrepreneurs. He has developed several Ornamental fish culture entrepreneurs in Tamilnadu through his intensive training programmes.

Dr S Felix, Ph.D., Vice Chancellor, Tamil Nadu Dr. J. Jayalalithaa Fisheries University, Nagapattinam is having experience of more than 35 years in the area of fisheries and aquaculture in teaching, research, extension and administration. He has organized more than 20 Nos. of seminar/conference/symposium at National and International levels. Currently he is also the President (2018-19) of World Aquaculture Society, Asian Pacific Chapter and Chairman, ICAR's BSMA Committee (2018-19). He has more than 60 research papers published in National and International journals. He has authored more than 10 books and 25 manuals. He is specialized in the area of advanced systems in aquaculture and aquariculture such as raceway, biofloc technology, RAS, etc,. He has operated around 20 externally funded research projects.

Aquatic Food Safety and Quality Management

by

F Parthiban

S Balasundari

B Ahilan

S Felix

Tamil Nadu Dr. M.G.R. Fisheries College & Research Institute
Tamil Nadu Dr. J. Jayalalithaa Fisheries University
Ponneri, Tiruvallur District, Tamil Nadu

DAYA PUBLISHING HOUSE®
A Division of
ASTRAL INTERNATIONAL PVT. LTD.
New Delhi – 110 002

ISBN: 9789388173193 (Int. Edition)

Published by : **Daya Publishing House®**
A Division of
Astral International Pvt. Ltd.
– ISO 9001:2015 Certified Company –
4736/23, Ansari Road, Darya Ganj
New Delhi-110 002
Ph. 011-43549197, 23278134
E-mail: info@astralint.com
Website: www.astralint.com

Laser Typesetting : **Classic Computer Services,** Delhi - 110 035

Printed at : **Neelam Graphics, Delhi - 110007**

डॉ. जे. के. जेना
उप महानिदेशक (मत्स्य विज्ञान)

Dr. J. K. Jena
Deputy Director General (Fisheries Science)

भारतीय कृषि अनुसंधान परिषद
कृषि अनुसंधान भवन-II, पूसा, नई दिल्ली 110 012

INDIAN COUNCIL OF AGRICULTURAL RESEARCH
KRISHI ANUSANDHAN BHAVAN-II, PUSA, NEW DELHI - 110 012

Ph. : 91-11-25846738 (O), Fax : 91-11-25841955
E-mail: ddgfs.icar@gov.in

Foreword

The health benefits associated with fish consumption have resulted a favourable consumer image for fish products. As a nation changes gear into a high state of development, new requirements and needs in life, especially in the food style emerge. Inevitably, there occurs a higher demand for quality and nutrition. And obviously people look forward to quality and safety in processed foods. Such requirement lead to adoption of international quality yardsticks such as the EU Standards, HACCP, ISO-9000 and ISO-2200 by the fish processing establishments. Therefore, it becomes apparent that care in processing has to start right from catching the fish, its subsequent handling and processing to get safe products for consumption.

Quality control, in the case of seafoods, means all the steps taken between harvesting and retail trade to protect the quality of the final product. Inspection consists of monitoring which is necessary to measure the effectiveness of the quality control procedure and also those official devices which are used to protect the consumer and facilitate trade.

This book entitled ***'Aquatic Food Safety and Quality Management'*** authored by ***Drs. F. Parthiban, S. Balasundari, B. Ahilan*** and ***S. Felix*** provides all necessary information on quality and quality parameters of fish and fishery products; quality and safety problem associated with seafood industry and their assessment; various standards applicable for export of fish and fishery products; methodologies for treatment of effluent water of seafood industry; and importance of overall food safety and quality system through adequate and appropriate understanding of the quality and safety parameters by the stakeholders in the food chain.

I appreciate the effort of the authors in bringing out this useful publication.

(J.K. Jena)

Message

In order to ensure the seafood safety and quality, the Food Safety Management System (FSMS) should enforce among the seafood processing units in the country. Hazard Analysis Critical Control Point (HACCP) and prerequisite programs are the most modern seafood quality management system introduced in India as recommended by importing countries like EU and USA. The HACCP system is based on the concept that various types of hazard can occur at various points in seafood processing and can be analyzed to locate critical control points and appropriate corrective measures can be taken to control the hazard for the purpose of production of safe product for human consumption.

TNJFU has been conducting various training program for implementation of HACCP in Indian seafood industries along with technology transfer. This book gives the occurrence of chemical, physical and biological hazards in the food processing line. It also gives practical approach of applying seven principles of HACCP, GMP, GHP for the production of safe and quality products. I hope this book will be great use to students and all experts in production, processing, storage, transportation and marketing of all food products particularly fish and fishery product

S. Felix
Vice Chancellor
Tamil Nadu Dr. J. Jayalalithaa Fisheries University

Preface

India exported 11,34,948 MT of seafood worth an all-time high of US$ 5.78 billion (Rs 37,870.90 crore) in 2016-17 as against 9,45,892 tons and 4.69 billion dollars a year earlier, with USA and South East Asia continuing to be the major importers while the demand from the European Union (EU) grew substantially during the period. Food safety is a concern facing the seafood industry today. Increased production of L. Vannamei, diversification of aquaculture species, sustained measures to ensure quality, and increase in infrastructure facilities for production of value added products were largely responsible for India's positive growth in exports of seafood. Besides frozen shrimp and frozen fish, India's other major seafood product was frozen squid. Export rejections from India have been on the higher side; primarily due to lack of proper infrastructure facilities with regard to adequate quality check at various levels. Shrimp farmers who procure the brood-stock for cultivation need to ensure the stock procured is free from viruses/contamination and the level of antibiotic in the water (where shrimps are cultivated) is within the permissible range. The second level of quality check takes place before the shrimps enter the processing stage and finally before the consignment is exported. There are extensive coverage in daily press on the food safety issues such as Salmonella, *Vibrio cholerae, Listeria monocytogenes,* transfer of antibiotic resistant microorganisms, quarantine issue related to viral pathogens, etc that threatening the consumers. It is very essential to create awareness among the Fisheries stakeholder with adequate knowledge on the above said aspects to tackle such food safety issues. This text book Aquatic Food Safety and Quality management provides detailed information on Quality Problems in Seafood Industry, Quality Assessment of Fish and Fishery Products, Concept of Quality

Management, Process Water Quality and Waste Water Treatment. The contents of this text book will serve as an excellent study reference material for students, researchers and entrepreneurs in the field of fish quality and safety.

Authors

Contents

	Foreword	*v*
	Message	*vii*
	Preface	*ix*
1.	**Introduction**	1
	1.1 Quality; 1.2 Quality Control; 1.3 Uniqueness of Seafood Quality Control; 1.4 Quality Assurance/Quality Management; 1.5 Intrinsic Quality; 1.5.1 Species; 1.5.2 Size; 1.5.3 Sex; 1.5.4 Condition and Composition; 1.5.5 Parasites and other Organisms; 1.5.6 Naturally Toxic Fish; 1.5.7 Contamination with Pollutants; 1.5.8 Occasional Peculiarities; 1.6 Extrinsic Quality; 1.6.1 Quality Deterioration and Extrinsic Quality Defects in Fish (Raw Material); 1.7 Fish Spoilage and Quality Assessment; 1.7.1 Enzymic Action; 1.7.2 Microbial Spoilage; 1.7.3 Non-enzymatic Deteriorations;	
2.	**Quality Problems in Seafood Industry**	13
	2.1 Poor Quality Raw Materials; 2.2 Spoilages and Quality Indices in Raw Fish; 2.3 Microbiology of Finfish; 2.4 Recognised Specific Spoilage Organisms (SSOs); 2.5 Mechanism of Microbial Spoilage; 2.5.1 Utilisation of Readily Utilisable Simple Substances; 2.5.2 Formation of Volatile Sulphur Compounds; 2.5.3 Hydrolysis of Complex Tissue Components into Simpler Substances and their Utilisation; 2.5.4 Breakdown of Nucleotides; 2.6 Spoilages and Quality Indices in Frozen Fish; 2.6.1 Freezing Process; 2.6.2 Frozen Storage; 2.7 Spoilages and Quality Indices in Canned and Retort Pouch Processed Fish; 2.7.1 Bacterial Problems in Canned Products; 2.8	

Defects in Canned Fishery Products; a. Blue/Black Discolouration in Canned Shrimp and Crab Meat; b. Green Discoloration; c. Brown Discolouration; d. Underweight in Canned Shrimps; e. Excessive Water Content in Oil Packs; f. High Sand Content in Canned Mussels and Clams; g. Struvite Formation; h. Honey-combing; i. Stack Burning; j. Perforation and Corrosion; k. Gaseous Spoilage; l. Non-Gaseous Spoilage; m. Mush; n. Curd and Adhesion; 2.9 Spoilages and Quality Indices in Cured Fish (Salted, Dried and Smoked); a. Pink Spoilage; b. Dun; c. Rancidity; d. Insect Infestation; e. Fragmentation; 2.10 Quality Control and Safety Aspects of Cured Fish; 2.11 Spoilages and Quality Indices in Surimi; 2.11.1 Quality Assurance in Surimi Processing; 2.11.2 Quality Evaluation of Surimi; 2.12 Quality Defects in Coated Fishery Products; i. Shelling; ii. Blow off; iii. Poor Adhesion; iv. Gummy Interface.

3. Quality Assessment of Fish and Fishery Products **36**

3.1 Methods of Quality Assessment; 3.2 Organoleptic/Sensory–Subjective and Objective; a. Appearance; b. Flavour; c. Taste; d. Odour; e. Texture; Interaction of Taste, Odour, Texture and Flavour; Sensory Testing Methods; 3.3 Selection of Test Subjects; i) Discrimination Tests; ii) Preference Tests; iii. Training of Panel Members; iv. Presentation of samples; 3.4 Sensory Testing Methods; 3.4.1 Preference Tests; 3.4.2 Discriminatory/Difference Tests; 3.4.3 Descriptive Sensory Analysis; 3.5 Sensory Evaluation of Fresh Fish; 3.6 Changes during Spoilage; i. Condition of the Skin (Sight and Touch); ii. Appearance of the Eye (Sight); iii. Condition of the Gills (Sight and Smell); iv. Condition of the Flesh (Touch and Sight); 3.7 Assessing Quality; 3.7.1 Raw Fish; 3.7.2 Cooked Fish ; 3.8 Secondary Characteristics; 3.9 Physical (Instrumental) Method for Assessing Seafood Quality; a. Freshness Meters; b. Texture Measurement; c. Electronic Nose; 3.10 Chemical/Biochemical Quality Evaluation of Seafood; 3.10.1 Volatile BNases; 3.10.2 Nucleotide Based Method; 3.10.3 Rancidity Tests; 3.10.4 Ammonia; 3.10.5 Biogenic Amines as an Index of Spoilage; 3.10.6 Indole as an Index of Quality; 3.11 Microbiological Quality Evaluation of Seafood.

4. Concept of Quality Management **71**

4.1 Traditional Quality Control; 4.2 Modern Safety and Quality Assurance Methods and Systems; 4.3 Methods to Manage Quality and Safety; i. Good Hygienic Practices/Good Manufacturing Practices; ii. Hazard Analysis Critical Control Point; iii. Quality Control (QC); iv. Quality Qssurance/Quality Management; v. Quality Systems; vi. Total Quality Management (TQM); vii. Good Manufacturing Practices in Fish Handling; 4.4 Food Safety with Special Reference to HACCP System; 4.4.1 The HACCP Concept; 4.4.2 Advantages of HACCP; 4.4.3 Applications of HACCP; 7. Determine Critical Control Points; 8. Establish Critical Limits for each CCPs; 9. Establish Monitoring System for each CCPs; 10. Establish Corrective Action; 11. Establish Verification

Procedures; 12. Establish Documentation and Record Keeping; HACCP Study Tips; Requirements for a Successful HACCP System; Current Problems; 4.5 Strategies for Prevention of Food Borne Diseases; 4.6 Food Safety Management System; 4.6.1 General Requirements; 4.6.2 Documentation Requirements; 4.7 Quality Standards; 4.7.1 Safety Standards; 4.7.2 Composition Standards; 4.8 National Standards; 4.8.1 Bureau of Indian Standards; 4.9 Export Inspection Council of India; 4.9.1 Inspection and Certification; 4.9.2 Fish Inspection Initatives and Resource Utilisation; 4.10 Export Procedures; 4.11 Export Documentation; 4.12 Food Safety and Standards Authority of India (FSSAI); 4.13 International Standards; i. ISO 9000 Series of Quality Systems and Total Quality Management; ii. ISO 9000 Systems; iii. Outstanding Features; iv. Advantages; v. ISO 9000 Requires a Documented Quality System Comprising of; vi. Assessment; vii. Benefits of ISO 9000; viii. Elements of Quality Management System; ix. Registration to ISO 9000 Standards; 4.14 USFDA; 4.15 Codex; a. The Codex Alimentarius Commission (Codex Standards).

5. Process Water Quality and Wastewater Treatment **129**

5.1 Quality Tolerances for Water for Processed Food Industry; 5.1.1 Tolerances for Taste; 5.1.2 Bacteriological Tolerances; 5.1.3 Physical and Chemical Tolerances; 5.1.4 Tolerances for Radioactivity; 5.1.5 Additional Tolerances for Specific Operations; 5.1.6 Additional Tolerances for Individual Food Industries; 5.2 Quality Tolerances in the Council Directive Relating to the Quality of Water Intended for Human Consumption (80/778/ EEC); 5.2.1 Organoleptic Parameters; 5.2.2 Physio-chemical Parameters (In relation to water's natural structure); 5.2.3 Parameters Concerning Substances Undesirable in Excessive Amounts; 5.2.4 Parameters Concerning Toxic Substances; 5.2.5 Microbiological Parameters; 5.2.6 Minimum Required Concentration for Softened Water Intended for Human Consumption; Conclusion; 5.3 Fish Plant Sanitation; i. Prerequisites to HACCP; ii. Good Manufacturing Practices (GMP); iii. Good Hygienic Practices (GHP); iv. Various Definitions of GHP or Prerequisite; 5.4 SSOP – Sanitation Standard Operating Procedures; The Processing Plant; 5.5 Disinfectants; i. Food Contact Surfaces are; ii. Sanitation; iii. Cleaning; iv. Cleaning Precedes Disinfection; 5.6 Detergents and Cleaning Schedule; 5.6.1 Terminology of Cleaning Agents used in Processing Detergents; 5.6.2 Rinse off the Detergent; Disinfectants; Sanitisers; Applications Used in the Food Industry; Sterilisation; Important Note; 5.7 Cleaning in Place; Importantance; 5.8 Waste Management in Fish Processing Industries; 5.9 Physico-chemical Parameters; 5.9.1 pH; 5.9.2 Solids Content; 5.9.3 Temperature; 5.9.4 Odour; 5.10 Organic Content; 5.10.2 Chemical Oxygen Demand; 5.10.3 Nitrogen and Phosphorous; 5.10.4 Characteristics of Fish Processing Wastewaters; 5.10.5 Sampling; 5.10.6 Discharge Limits; 5.11 Primary Treatment; 5.11.1

Screening; 5.11.2 Sedimentation; 5.11.3 Separation of Oil and Grease; 5.11.4 Flotation; 5.12 Biological treatment; 5.12.1 Aerobic Processes; 5.12.2 Anaerobic Treatment; 5.13 Physico-chemical Treatments; i. Coagulation - Flocculation; 5.14 Disinfection; i. Chlorination; ii. Ozonation; 5.15 Sludge Treatment and Disposal; 5.15.1 Sludge Transportation to Treatment Facilities; 5.15.2 Sludge Digestion; 5.15.3 Land Disposal of Sludges; 5.15.4 Sludge Disinfection.

Chapter 1

Introduction

The health benefits associated with fish consumption have resulted in a favourable consumer image for fish products. As a nation changes gear into a high state of development, new requirements and needs in life, especially in the food style are emerging. Inevitably, there occurs a higher demand in quality and nutrition. And obviously people look forward to quality and safety in processed foods. Upgraded international quality yardsticks such as the EU Standards, requirements of HACCP ISO-9000 and ISO-2200 series are internationally being adopted in fish processing establishments. Therefore it becomes apparent that care in processing has to start right from catching the fish, its subsequent handling and processing to get safe products for consumption.

1.1 Quality

Quality is generally considered as the degree of excellence. Quality is often related to the price at which a commodity is purchased or the purpose for which it is to be used. In relation to seafoods, quality is the sum total of its composition, nutritive value, degree of freshness, physical damage, deterioration while handling, processing, storage, distribution and marketing, hazards to health, satisfaction on eating yield and profitability to the producer and middleman. In short, fish quality means all those attributes which a fish consumer considers to be present individually or collectively in the item.

1.2 Quality Control

Quality Control (QC) can be defined as the operational techniques and activities that are used to fulfil quality requirements. Quality control, in the case of seafoods, means all the steps taken between harvesting and retail trade to protect the quality of the final product. Inspection consists of monitoring which is necessary to measure the effectiveness of the quality control procedure and also those official devices which are used to protect the consumer and facilitate trade.

1.3 Uniqueness of Seafood Quality Control

Quality control of seafoods differs a lot from that of any other food. Fruits, vegetables, *etc.* are harvested under ideal conditions i.e. the right type of food is harvested at the right time and at the right place. This means that it is possible even to select the right species, to rear them to the desired level of growth and harvest them in a pre-determined place as per pre-determined schedule. In the case of seafoods, the harvesting is from the sea, rivers and other water areas from the stocks of unknown identity as to age, sex *etc.* under the most difficult conditions. Hence, the intrinsic quality varies in all possible combinations. Another phenomen on which complicates the quality control of seafoods is the unpredictability of catch. At times, the catches may exceed the handling capacity even ice may not be available in sufficient quantities. Sometimes, catches are poor and the processing units have to remain idle for weeks together. Another peculiarity of tropical fisheries is the presence of large number of species in limited quantities making quality control more difficult.

1.4 Quality Assurance/Quality Management

This can be defined as all the activities and functions concerned with the attainment of quality in a company. In a total system, this would include the technical,managerial and environmental aspects as alluded to above. The best known of the quality assurance standards is ISO 9000 and for environmental management, ISO 14000.

The term quality management is often used interchangeably with quality assurance. In the sea food industry, the term quality management has been used to focus mostly on the management of the technical aspects of quality in a company, for instance, the Canadian Quality Management Programme which is based on HACCP but covers other technical issues such as labelling.

1.5 Intrinsic Quality

It is inherent quality. Several factors affect the intrinsic quality. They are species. Size, sex, condition and compositions, parasites and other organisms, naturally toxic fish, contamination with pollutants and occasional peculiarities.

1.5.1 Species

Certain fishes are generally costly because they are rated as **'good quality'** fishes. In India, seerfish and pomfret are costly and hence, rated as **good quality fish**. In this case, the quality is related to species. Hence, species identification assumes importance where the external features of the fish are destroyed while processing as in the case of frozen fillets or minced meat.

1.5.2 Size

Generally large specimens of a given species fetch better price. Large sized shrimps, crabs and lobsters fetch higher price in the markets. Processors place higher value on large specimens because of higher percentage yield of edible material, lower handling costs per unit weight, better keeping qualities and yielding of more uniform products. But for certain purposes, the biggest size is not the optimum as in the case of canned shrimps because of difficulties in controlling fill weights. Control of size is to some extent exercised by choosing the fishing grounds, seasons or methods. Manual or mechanical sorting is also possible after catch.

1.5.3 Sex

Preference of fish based on sex is known to exist, but there is no documentary evidence of this, where Indian waters fish species are concerned. The females of certain species such as sturgeon are more valued in Europe probably for their roe or caviar.

1.5.4 Condition and Composition

Condition refers to the biochemical and physiological state. In certain seasons, fish is said to be in good condition and at other times in poor condition. Fish in poor condition is known to have more water content and less fat content. This is usually so after spawning. Sardines and mackerels are known to show wide variation in their fat content depending on season. Sardines in October, November and December have very high fat content and are preferred by May during this season. Feedy fish are known to be prone to belly bursting and hence they cannot be kept for long. Glycogen content in certain shellfishes varies with season. When glycogen is lower, the amount of lactic acid formed is also lower and hence the pH higher. This means that such fish will subject to microbial attack earlier. Crabs are known to differ in their meat content dependent on full moon or new moon periods. During new moon period, their meat content is higher. A special problem in certain fishes is 'chalkiness' *i.e.* the fish flesh becomes whiter during certain seasons and is related to the more acidic pH attained. Considering many of these points, Regulations exist in some nations on fishing seasons. Low post-mortem pH also leads to 'gaping' *i.e.* the tendency of fish fillets to split into fissures and holes.

1.5.5 Parasites and other Organisms

A parasite is an organism living on or inside another and depending upon it for some of its vital needs, particularly nutrients. Incidence of parasites in fish is another quality factor. These parasites are not easily visible and can cause difficulties when discovered by the consumer. Only few of these parasites are harmful. Many of these parasites are located in the head, viscera, *etc.* which are not consumed. Though thorough cooking kills all parasites and renders them completely harmless, these parasites cause health problems where fish is consumed raw. Some parasites are seen in tuna. The parasites in fish mostly belong protozoa, flat-worms (platyhelminths), round worms (nematodes) and certain crustacea. Becteria and fungi also cause diseases in fish. Protozoan such as myxosporidian infection may cause softening of fish *e.g. Chloromyscium thyrsites.* Many flat worms are free living but flukes (trematodes) and tape worms (cestodes) are parasites. Diseases in human beings are caused by consuming the cysts of two types of flat worms, the lung fluke (*Paragonimus* sp.) and the broad tapeworm (*Diphyllobothrium latum*). Round worms (Nematodes) may be found in the gut and viscera of fish or encysted in the flesh. The cod worm (*Porrocaecum decipiens*) is a well – known example and other example is *Anisakis* sp., which is fairly common in squids and herrings. The occurrence of crustacean, fungi and bacteria is not a major problem in proecessing. Crustacean contains many parasitic forms, mostly belonging to the sub – group known as copepods. The sea louse, *Lepeophtheirus* sp.that is found on fresh run salmon and sea frout, is an obvious crustacean. Fungal and bacterial infections are widespread and can be of particular importance in fish farming and so only few are of importance for the fish processors. An example of a fungal disease is the condition known in the UK as 'greasy haddock'. It is caused by *Ichthyospor idium hoferi.* The flesh of infected specimens is soft and has a some what sweetish, slightly sickly smell. When such fish is smoke, white spots are seen in the flesh. Lesions, nodules (small lump or swelling) and pustule (pimple or blister) areas found in the commercially caught fish may be attributable to bacteria, mainly *Aeromonas* sp.

1.5.6 Naturally Toxic Fish

Some fishes are naturally toxic and cause injury to health or even death. Fish and shellfish poisonings are due to toxins in the flesh of animals either intrinsically present or derived from the food they consume. This is true for puffer fish poisoning (caused by tetradotoxin), ciguatera (caused by cigua toxin and possibly other toxins) and paralytic shellfish poisoning (caused by saxi toxin and other toxins). Quality control personnel should be able to identify such species. Ciguatera poisoning occurs by consuming certain carnivorous fish such as barracudas, snappers, *etc.* harvested from shallow waters in or near tropical and subtropical coral reefs at certain seasons. The ciguatoxin (CTX) is responsible for the poisoning and the principal source is the benthic dinoflagellate, *Gambierdiscus toxicus.*It is rarely fatal. Cooking does not destroy the poison. Puffer fish poisoning is by consuming the puffer fish flesh contaminated with viscera. Mortality rate is 50 per cent and largest incidence is reported from Japan. Tetradotoxin (TTX) is caused by the bacteria, *Vibrio alginolyticus.* Paralytic shellfish poisoning is caused by eating certain molluscs particularly mussels and clams. The toxicity is related to the occurrence of dinoflagellates such as *Gymnodium* sp. or *Ptychodiscus* sp. in the harvesting water. When these organisms are too much, the water assumes a reddish tinge and it is called as red tide phenomenon. The only protective measure is not to collect shellfishes from such waters. Paralytic shellfish poisoning (PSP) is also called either as mussel poison or mytilo toxin/clam poison or saxi toxin. Common sources of PSP are *Gonyaulax catenella*, *Alexandrium* and *Pyrodinium* sp. The toxin responsible for diarrhetic shellfish poisoning (DSP) is okadic acid (OA) dinophysis toxin and the common source of neurotoxic shellfish poisoning (NSP) is also called as oyster or asari poisoning and its common source is Prorocentrum sp. Amnesic shellfish poisoning (ASP) affects the brain leading to the loss of memory and the toxin responsible is domoic acid. ASP is the only shellfish poison produced by a diatom. Erythematous shellfish poisoning (ESP) affects the blood circulatory system by specifically attacking the red blood cells.

1.5.7 Contamination with Pollutants

Contamination of fish with pollutants is another problem, which an ordinary fish consumer is unable to identify. The main classes of pollutants, which need to be considered, are metals and other elements, chlorinated hydrocarbons, mineral oils, radioactive isotopes and microorganisms. A large number of potentially harmful heavy metals and elements that include mercury (0.5 to 1 ppm), cadmium (2 ppm), lead (1.5 ppm), selenium (1ppm) and arsenic (1 ppm) are known pollutants. The important persistent organic chemicals are a group of chlorinated hydrocarbons including the insecticides/pesticides, DDT, aldrin,dieldrin, benzene hexachloride (BHC or lindane) and polychlorinated biphenyls (PCB's). The US Food and Drug Administration have imposed maximum permitted concentrations for DDT and dieldrin in most fish as 5.0 and 0.30 mg per kg wet weight, respectively. The residues of antibiotics such as chloramphenicol, nitrofurans, tetracyclines, *etc.* are in the flesh of fish and shellfish also dangerous to human beings. Very occasionally catches become grossly contaminated with chemicals, especially mineral oils resulting from accidental (during offshore oil exploration and oil tanker movements) or other kinds of large – scale release. Evidence of such contamination is apparent in tainted odour or flavour of the fish. Exposure to high energy radiation originating from radioactive isotopes can be extremely injurious to health. Naturally occurring radioactive isotopes occur in fish, as in other foods, but at very low concentrations not considered to have a significant effect on health. In some areas, artificial isotopes have been released in very small amounts into the environment.

Although most of the microorganisms present on the outer surface, gills and in the viscera of fish caught in unpolluted water are harmless to man, food poisoning can be caused by two microorganisms that may occasionally occur. Pollution of fish by raw or inadequately treated sewage contains two types of faecal microorganisms such as bacteria and single member of the virus family. The bacteria include the large group of Salmonella and different members of which cause food poisoning, typhoid and paratyphoid and Shigella, which cause dysentery. The only virus known to be incriminated is that responsible for the severely disabling disease infectious hepatitis. These microorganisms are transferred via seawater and detritus to fish including crustaceans and molluscans living within several kilometres of sewage out-falls or sewage polluted rivers. No sewage microorganisms are normally detectable in the open sea. The molluscans harvested from the sewage-polluted areas cause health hazard if they are consumed raw, as they concentrate microorganisms because they are filter feeders. A large number of different types of faecal microorganisms and of pathogens are potentially present in any given sample but is far too expensive and consuming to determine more than one or two of them. It is therefore a standard practice to select what is known as an indicator microorganism which is invariably present in sewage and whose behaviour in the environment and in the molluscans is to all intents and purpose indistinguishable from that of the pathogens. The indicator organisms now being generally determined are either a group known as faecal coliforms or the overwhelmingly predominant number of the group, *Escherichia coli*. Determinations of their indicator give very similar results. Limits have been specified for these contaminants by national as well as international quality control organizations.

1.5.8 Occasional Peculiarities

Peculiarities like tumors, ulcers, nodules, abnormal colouration, abnormal odours, *etc.* are also rarely observed in fish. Physical damage to food fish caused by predators such as seal on salmon, shark on many species before capture is occasionally found. But, they do not pose a serious problem in processing. The flesh of cod and other gadoids is on certain occasions found to be in various shades of pink colour. This is due to the presence of the natural red carotenoid pigments such as astaxanthin and zeaxanthin believed to be derived from unusual feed or some metabolic disturbance. On oxidation, these red pigments turn into yellow colour. The most well- known of the peculiar odours and flavours is reported in cod and other gadoids, mackerel and chum salmon. It is variously described as 'black berry, weedy, petrol, diesel, iodine and sulphide'. The substance responsible for this has been identified as dimethylsuphide (DMS). This does not occur normally in most fish, but may occur as result of feeding to certain organisms. Species of planktonic bivalve molluscs known as pteropods have been implicated, particularly **Spiratella helicina or Limacine helicina**. These pteropods contain dimethyl –b-propiothetin which is converted in the fish to DMS.

Freshwater fish including tilapia occasionally suffer from an earthy or muddy odour and flavour, which can reduce consumer acceptance. Earthy odour occurs particularly in fish caught in waters harbouring high concentrations of certain algae particularly blue green algae or microorganisms belonging to Actinomycetes that they have as imilar odour. The earthy odour is probably caused by the presence of geosmin and related substances originationg in algae or microorganisms present in the water. An iodoform like odour in shrimps is fairly common and can be objectionable in high concentration. High concentrations of geosmin (trans, 1-10-dimethyl-1-9 decalol), a musty odorous compound, are found in the tail muscle of the shrimp grown in coastal culture ponds. It is responsible for the earthy – musty flavour in pond cultured penaeid shrimps.

1.6 Extrinsic Quality

It is due to the effects of treatment the fish receives ofter capture till it reaches the consumer. The important extrinsic quality factors are degree of freshness, conformity to the declared mode of presentation, weight, size, ingredients and food additives, acceptability of the processing methods and suitability of containers and packaging methods. Special significance is attached to the cleanliness and sanitary conditions under which the products are handled and processed as evidenced by the degree of contamination with pathogenic and faecal indicator organisms.

1.6.1 Quality Deterioration and Extrinsic Quality Defects in Fish (Raw Material)

Quality deterioration means that those natural processes of quality reduction, which occur after harvesting and that are quite independent of deliberater intervention of human beings, whereas extrinsic defects are quality reductions in the post - harvested material caused by deliberate or accidental actions of human beings. There is a possibility of a considerable degree of planned control over both quality deterioration and quality defects. In the fish industry, it is sometimes impossible to draw a clear distinction between raw material and products, because, what is raw material to a processor may be finished product to a retailer. The raw material is considered as chilled or unchilled whole fish and shellfish, gutted or ungutted fish, and unpeeled or unshucked raw shellfish, and in every case it refers to raw uncooked material. The condition of such raw material is basic to the quality of many products. The causes of quality deterioration are broadly classified in to microbiological and non - microbiological.

1.7 Fish Spoilage and Quality Assessment

Fish is a highly perishable commodity. Spoilage of fish begins as soon as the fish dies. In tropical conditions, fish spoils quite rapidly, within a few hours of landing, if not properly cooled. The spoilage rate of fishmay be reduced by good handling practices and effective temperature control from the very beginning.

In raw fish, spoilage takes place mainly due to three reasons *viz.*,

1. Enzymic action
2. Microbial action
3. Non enzymatic Chemical action

1.7.1 Enzymic Action

Enzymes and bacteria do not cause any deteriorative changes in the living cell bacause of the natural defensive mechanism. In dead fish, enzymes become involved in autolytic changes and bacteria can invade the fish muscle and proliferate there. The fish gut is rich in proteolytic enzymes and in dead fish it digests the gut and belly region making the fish very soft. Bacteria that are present on the surface, gills and gut ofthe fish invade the dead fish, decompose the tissue and bring about undesirable changes. Off odours and off flavours, slime, gas production, discolouration and soft texture are the obvious signs of spoilage. Components involved in spoilage process are protein, lipids, carbohydrates, nucleotides and other non-protein nitrogen compounds. The rate of spoilage is temperature dependent and lowering the temperature will reduce the rate of spoilage.

1.7.1.1 Enzymatic Deteriorations/Autolytic Spoilage

Autolytic spoilage is responsible for early loss of quality of freshfish. The first enzymatic change in fish muscle is the gradual hydrolysis of glycogen to lactic acid which is known as glycolysis.

a. Glycolysis

After death, the blood circulation ceases and the cells are no longer supplied with oxygen and hence glycogen will not be converted into carbondioxide and water unlike in the case of living cells. In the post-mortem period, glycolysis proceeds via the anaerobic pathway where the end product is lactic acid. As lactic acid accumulates, the pH of the muscle falls. In fish, glycolysis will continue until the supply of glycogen is completely used up. In general, fish muscle contains a relatively low amount of glycogen compared with mammalian muscle, and the final post-mortem pH is consequently higher. This makes the fish meat more susceptible to microbial attack. However, there is great variation in the glycogen contents of different species; for example tuna has levels comparable with those found in mammals. Rested fish contains more glycogen than exhausted fish and well fed fish more than starved fish. Within the fish, glycogen is more concentrated in the dark muscle than in the white muscle. In stressed fish, glycogen is rapidly depleted. The lactic acid formed lowers the pH from 7.0 - 7.2 to around 6.2-6.5. In some species the final pH will be 5.8 - 5.6. The decline in pH is accompanied by the natural postmortem stiffening called rigor mortis. Rigor mortis begins 0 - 8 hrs after death and lasts for 10 - 120 hrs. It is depending on exhaustion, temperature, handling (physical damage) and sizeof fish. The rigor is resolved after this stage.

b. Flavour Changes in Fish (Nucleotide degradation)

The most significant enzyme deteriorations are those that affect flavour.The nucleotide degradation in fish muscle produces many flavour bearing compounds. These compounds are formed by the splitting of ATP (Adenosine Triphosphate) by a series of dephosphorylation and deamination reactions.

ATP $\rightarrow$ ADP $\rightarrow$ AMP $\rightarrow$ IMP $\rightarrow$ Inosine $\rightarrow$ Ribose $\rightarrow$ Hypoxanthine.

Progressively it is hydrolysed to ADP (Adenosine Diphosphate), AMP (Adenosine Monophosphate) and to IMP (inosine monophosphate) and ammonia. At ambient temperature, ATP breakdown occurs very rapidly and IMP accumulates in the fish tissue. In fresh fish, the level of IMP is very high and it imparts a desirable sweet, meaty and characteristic flavour to fish. As autolysis proceeds further, the level of IMP decreases and neutraltasting inosine or bitter tasting hypoxanthine accumulates in the tissue. As a result, fish becomes more insipid. Some of these compounds increases with time and have been used as indices of fish freshness.

c. Belly Bursting

Enzymic spoilage causes belly bursting in fish, especially during a period of high food intake. These fishes will have a large content ofdigestive enzymes in the digestive tract. Such fish will degrade quickly and spoil easily soon after they are caught. In the dissolved gut components,bacteria proliferate and produce gases such as CO_2, and H_2. This gas production leads to belly bursting after short storage period. Keeping the fish live for some time will retard this.

d. Colour Changes in Fish

Colour is an important factor in seafood quality. Colour change in seafoods is caused by enzymic or non-enzymic action such as fat oxidation or by pigments. Colour change in seafood is an indication of spoilage. The important discolouration in seafoods caused by enzymes or fatoxidation is illustrated below:

(i) Shrimp

The development of black spot in shrimp is due to the presence of an enzyme, polyphenol oxidase (PPO). The black spot (melanin) is formed by the oxidative reaction of tyrosinase on tyrosine. Different species of shrimp undergo postmortem discolouration to different extents. The pigment is formed on the internal shell surfaces, or in advanced stages, on the underlying shrimp meat. This makes the shrimp unattractive for marketing.Ths shrimp phenol oxidase differs from that found in mussel or lobster in that it is not activated by trypsin. Sulphite preservatives can be used to prevent black discolouration resulting from phenolase reaction. Dipping shrimp in 0.2–0.5 per cent sodium metabisulphite for one minute is usually adopted in industrial practice.

(ii) Lobster

Black spot occurrence during icing and frozen storage of lobster is a serious problem which has resulted in significant commerical losses.Tyrosinase activity was identified in the blood with lesser amounts in other tisses.Discolouration at the butt of the tail, black spot between the segments and other deteriorations are more pronounced in moribund losters. The tendency for blackening is also influenced by moulting cycle.

(iii) Crab

Phenolases have been implicated in blue-black colouration of crab.The enzyme is present in the blood. Various additives have been shown to prevent blue discolouration (*e.g.*: organic acids, sodium bisulphite and EDTA). The haemocyanin induced blackening is also prominent in crab and is non-enzymic.

(iv) Yellowing of Fish Flesh

Frozen storage of some fish may result in yellowing of flesh below the skin. Freezing or other processes disrupt chromatophores and release carotenoids and their migration to the subcutaneous fat layer causes yellowing. Yellowing associated with lipid oxidation and carbonyl-amine reaction is observed during frozen storage.

(v) Brown Discolouration

Brown or yellow discolouration is caused by the reaction of protein or amino acids with product of lipid oxidation. Brown discolouration is observed in a variety of processed products including white pomfret, sardine, jack mackerel, salted shark, marine eel *etc.* Discolouration due to protein-lipid browning is greater in fatty fish than lean fish.

1.7.2 Microbial Spoilage

Fish spoilage is mainly due to the action of bacteria. Bacteria are present on the surface slime, skin, gills and intestine of fish. In dead fish, bacteria begin to invade the tissues causing spoilage and production of undesirable compounds. The type of bacteria on the fish produce very much low levels of this compound (0.1 - 0.5 per cent), while elasmobranchs (shark, rays *etc.*) and gadoids contain very high levels (1 - 1.5 per cent). Spoilage bacteria convert this substance into foul smelling trimethylamine (TMA). TMA is produced in fish muscle slowly at first then at a greater speed in fish stored at ambient temperature, in ice or in refrigerated seawater. The fishy odour is produced when it reacts with fat.

I) Histamine Formation

Microbial spoilage of fish produces the toxin, histamine in certain fishes. Histamine poisoning or scombroid fish poisoning is very frequent in many countries. Scombroid fishes and other dark muscle fishes contain high levels of free amino acid, histidine, in their muscle. During spoilage, histidine is converted into histamine by bacteria. Over 50 species including popular species such as tuna, bonito, mackerel, blue fish, dolphin fish (*Mahi mahi*), carangids, herring, sardine and anchovies have shown to be a potential threat of histamine poisoning. Histamine production increases with temperature and 37 °C is the optimum temperature for microbial activity. *Morganella morganii, Klebsiella pneumoniae* and *Hafnia alvei* are the main spoilage organisms producing histamine. Low temperature storage, right from catch, reduces histamine production.

II) Indole Production

Conversion of tryptophan to indole is another result of amino acid decomposition by bacteria. The FDA uses indole level along with sensory evaluation for measurement of shrimp decomposition.

III) Other Compounds Formed during Bacterial Spoilage

A number of other extractives are available in fish for bacterial action such as free amino acids, sugars, peptides, creatine, as well as lipid and proteins. Chemical examination of spoiling fish muscle has shown that organoleptically the most important constituents are the three volatile sulphur compounds such as hydrogen sulphide (H_2S), dimethylsulphide $(CHI)_2S$ and methylmercaptan (CH_3SH). Esters of lower fatty acids such as acetic, propionic, butyric and hexaenoic acids are also produced. Volatile sulphur compounds influence the organoleptic characters, especially odours, in spoiling fish. The overall qualitative chemical picture of spoiling fish is summarised below

Table 1.1: Chemical Compounds Produced by Bacterial Action

Substrate	*Compounds Produced by Bacterial Action*
Inosine	Hypoxanthine
Carbohydrate and Lactate	Acetic acid, CO, and H_2O
Methionine and cysteine	H_7S, CH_3SH and (CH_3)-,S
Tryptophan	Indole
Glycine	Esters of acetic, propionic, butyric and
Leucine and Serine	hexaenoic acids
Trimethylamine oxide	Trimethylamine
Urea	Ammonia
Lipids	Carbonyls
Proteins	Tyrosine, indole, skatole, putrescine, cadaverine
Histidine	Histamine

Some of the spoilage bacteria are proteolytic and undoubtedly contribute to the ammoniacal odour by producing ammonia from the protein breakdown.

IV) Pathogenic Bacteria

The pathogenic bacteria associated with seafoods are of two types:

a. Indigenous Bacteria

They are widely distributed in the aquatic environment. These pathogens occur in minimal numbers and are not a serious problem in fresh fish. However, their growth and multiplication in seafood is a serious problem and can cause illness. *e.g. Clostridium botulinum, Vibrio* sp., *Aeromonas* sp.

b. Non-indigenous Bacteria

They occur in seafood as a result of contamination. The source include polluted aquatic environment, sewage, excreta from animals, birds, human beings, workers handling the material as well as the surface and environment where the seafood is processed. Eg. *Salmonella sp., Shigella, E. coli* and *Staphylococcus aureus.*

1.7.3 Non-enzymatic Deteriorations

Fish lipid is characterised by a high level of polyunsaturated fatty acids (PUFA) and hence undergoes oxidative changes with fatty fish in particular, fat oxidation gives rise to problems such as rancid flavour and odour as well as discolouration. Lipid oxidation is by two processes:

1. Auto oxidation - action of O_2 on the unsaturated fatty acids and
2. Lipid hydrolysis - an enzymatic hydrolysis with free fatty acids (FFA).

Oxidative rancidity is of great concern in fatty fish storage. In pelagic species like sardine, mackerel and herring, rancidity has been detected during spoilage.

At first, hydroperoxides are formed, which further degrade to form aldehydes and ketones with typical rancid flavour. The oxidation is initiated and accelerated by heat, light (UV-radiation), presence of several organic or inorganic compounds (*e.g.* Cu and Fe), moisture content, large surface area, presence of air *etc.* Antioxidants such as a tocopherol, ascorbic acid, citric acid or carotenoids can inhibit oxidation.

Chapter 2

Quality Problems in Seafood Industry

Our seafood products can broadly be divided into two – Export products and Domestic products. The quality of the products in both these cases has to be good. Some of the most common quality problems usually encountered in these products are detailed below:

2.1 Poor Quality Raw Materials

We notice wide variations in the quality of raw material received in the seafood processing plant. This problem is dependent on size of the fishing craft, facility for icing and storage of fish on board the fishing vessel, facilities for further preservation and storage at the landing centre, sanitary conditions on board the vessel, pre-processing and processing centres *etc.* Attention is needed in the primary processing centre to see that:

1. Washing off filth materials before starting any processing
2. The beheading operation has been satisfactorily carried out
3. Peeling and deveining operations are complete
4. Broken pieces are separately kept/Segregation of broken and loose pieces
5. The raw material temperature is never allowed to exceed 2°C
6. Good quality water and ice are used for processing and
7. Strict hygienic and sanitary conditions are maintained the processing unit.

2.2 Spoilages and Quality Indices in Raw Fish

Bacterial spoilage is caused by the activities of microorganism associated with the fish. Bacterial spoilage of fish begins only after the completion of rigor mortis, which results in the release of end products of protein denaturation due to decrease in pH, which are utilizable by bacteria. Thus, prolonging rigor mortis helps to delay spoilage and thereby keeps fish fresh.

Spoilage of both marine and fresh water fish occurs in the same manner. Fish contain high levels of protein and non- protein nitrogenous constituents (16~20 per cent), lack carbohydrate(very less amount), and have varying amounts of fat depending on the species of fish. The non-protein nitrogenous compounds in fish include free aminoacids, volatile nitrogen bases- ammonia and trimethyl amine (TMA), creatine, taurine, betaines, uric acid, anserine, carnosine and histamine. Spoilage of fish begins from the surface, gill and intestine because of high bacterial load. From gills, intestine and surface microorganisms gradually migrate to adjacent tissue and cause spoilage. Spoilage organism first utilizes simpler compounds and later fish proteins releasing various off-odour compounds.

The microbial load and types of microorganisms associated with the freshly harvested fish directly reflects on the microflora of the immediate environment from which fish is harvested. Further, several microorganisms are added to the fish from harvesting gear and net, from the boat deck, fish contact surface, fish holds and fishermen. Once the fish dies, microorganisms present on the body surface, gills and the intestine start multiplying and increase in numbers causing deteriorative changes in fish. As the time lapses the spoilage proceeds faster making the fish unfit for human consumption. The spoiling fish is generally characterized by the loss of bright body coloration, fading of gill colour, sunken eyes and development of off odour metabolites of spoilage bacteria. The spoilage flora of fresh fish is generally dominated by the Gram negative bacteria. The effective way to prevent or delay spoilage of fresh fish is by reducing the activity of spoilage microorganisms which can be achieved by lowering the temperature of holding the harvested fish.

2.3 Microbiology of Finfish

The surface flesh of live, healthy fish is considered sterile and devoid of any bacteria or other microorganisms. On the contrary, as with other vertebrates, microorganisms colonise the skin, gills and the gastrointestinal tract of fish. The number and diversity of microbes associated with fish depend on the geographical location, the season and the method of harvest. In fact, standard culture-dependent methods can only recover between 1 to 10 per cent of total bacteria present in any given sample. The total number of bacteria ranges from 102 to 107 cfu/cm^2 on the skin surface and between 103 and 109cfu/g for both in the gills and intestines. Fish harvested from clean and cold waters have lower bacterial numbers than fish from eutrophic and/or warm waters. Also, fish from polluted warm waters contain up to 107cfu/cm^2. However, potential human pathogens may be present in both scenarios.

The autochthonous bacterial flora of fish is dominated by Gram-negative genera including: *Acinetobacter, Flavobacterium, Moraxella, Shewanella* and *Pseudomonas*. Members of the families *Vibrionaceae* (*Vibrio* and *Photobacterium*) and the *Aeromonadaceae* (*Aeromonas* spp.) are also common aquatic bacteria, and typical of the fish flora. Gram-positive organisms such as *Bacillus, Micrococcus, Clostridium, Lactobacillus* and coryne forms can also be found in varying proportions. It is crucial to mimic the environmental physico-chemical parameters when isolating bacteria from fish.

It is apparent from the above that there is potentially a very diverse range of organisms present on fish. However, numbers of pathogenic bacteria in raw fish tend to be low, and risk associated with the consumption of seafood is low. In addition, during storage indigenous spoilage bacteria tend to outgrow potential pathogenic bacteria. Shelf life depends on the initial microflora on the fish, potential contaminants added during handling and processing and conditions of storage.

2.4 Recognised Specific Spoilage Organisms (SSOs)

The degree of spoilage leading to sensory rejection of fish is partly dependent on the perception of the consumer. Not all the bacteria growing on a food will lead to the production of objectionable characteristics; a minority are often associated with the majority of the spoilage. The concept of specific spoilage organisms (SSOs) is not new; yoghurt spoilage by yeasts and clostridial spoilage of cheese are examples where it has been recognised for many years that a particular minority of the microbial flora present in the product is responsible for its spoilage. For fresh fish, realisation that the bulk of the microbial population on newly caught fish does not cause off-flavours and off-odours stems from work. The spoilage of a product may be strongly influenced by the conditions under which the product is held; therefore, the characterisation of spoilage of each product must be made before the identification of the responsible agent(s) can proceed. Bacteria identified as being associated with the spoilage process of fresh fish are as follows:

i. *Pseudomonas* spp.

The Pseudomonadaceae family represent a large and poorly defined group of microorganisms. The spoilage compounds associated with the growth of psychrotrophic *Pseudomonas* spp. on fish are diverse and in many cases species-specific. *Pseudomonas* spp. mediated spoilage is characterised by 'fruity', 'oniony' and'faecal' odours from the production of ketones, aldehydes, esters and non-hydrogen sulphide sulphur-containing compounds such as methyl sulphide. Members of the genus are able to produce pigments, and proteolytic and lipolytic enzymes that may affect the quality of fresh and, more especially, processed (*e.g.* frozen) fish products.

ii. *Shewanella putrefaciens*

Shewanella putrefaciens spoilage of fish is due to its biochemical action on muscle, *i.e.* its ability to reduce TMAO to TMA, produce hydrogen sulphide (H_2S)

from cysteine, form methylmercaptane (CH_3SH) and dimethylsuphide ($(CH_3)_2S$) from methionine and produce hypoxanthine (Hx) from inosine monophosphate (IMP) or inosine, plus other characteristic compounds of the species responsible for spoilage.

iii. *Photobacterium phosphoreum*

Recognised for some time as being present on spoiling fish, *Photobacterium phosphoreum* increased in notoriety when it was proposed that reduction of TMAO to TMA limited the shelf life of MAP cod fillets. Because no other TMAO-reducing bacteria were present in sufficient numbers to produce the quantities of TMA that were related to rejection, it was proposed that this organism, owing to its cell size and activity, was capable of being in a significant numerical minority on the MAP fish but still able to yield the majority of the TMA thought to be responsible for the rejection.

iv. *Brochothrix thermosphacta* and Lactic Acid Bacteria

B. thermosphacta is a well-characterised psychrophilic spoilage organism of meat. Growing evidence suggests a role for *B. thermosphacta* in the spoilage of some MAP fish. Recent studies have investigated the dominance of *B. thermosphacta* on spoiling fish in a 40 per cent $CO_{2,}$ 30 per cent N_2, 30 per cent O_2 MAP. Acetate production has been reported as a good indicator of spoilage by this organism. MAP studies have also demonstrated this organism's sensitivity to oxygen and have shown that they are also inhibited by high CO_2 concentration.

2.5 Mechanism of Microbial Spoilage

Microbes decompose and spoil fish in ways as given below:

1. Utilisation of readily utilisable simple substances such as non-protein nitrogenous (NPN) substances
2. Hydrolysis of complex tissue components (*e.g.* proteins) into simpler substances (*e.g.* amino acids) and their subsequent utilisation
3. Breakdown of nucleotides.

2.5.1 Utilisation of Readily Utilisable Simple Substances

The NPN compounds are readily utilisable simple substances consisting of TMAO, Urea and free amino acids.

i. Trimethylamine Oxide (TMAO) Degradation

TMAO is an osmoregulatory substance present in the tissue of marine finfish and shellfish. Negligible quantities of TMAO are present in freshwater fishes. The elasmobranchs contains the highest and the flat fishes the lowest TMAO. The muscle of sharks contains 750-1480 mg per cent TMAO. White fleshed demersal fishes contain more TMAO than dark fleshed pelagic fishes.

TMAO is subsequently degraded through reduction to trimethyamine, dimethyl amine, methyl amine and finally to ammonia. Reduction of TMAO involves reduction of one methyl group in each step with its liberation as methane.TMAO is reduced to TMA enzymatically by the enzyme TMAO reductase. This enzymatic reduction takes place by microbial enzyme as well as by tissue enzymes.The TMAO reduction is mainly associated with the genera of bacteria such as *Alteromonas, Photobacterium, Vibrio, Shewanella putrefacians, Aeromonas* and intestinal bacteria of Enterobacteriaceae. TMA is responsible for the typical fishy odour of the marine teleosts.The content of TMA in fish tissue is expressed as TMA-N and is used as an indicator of microbial spoilage. The level of TMA found in fresh fish rejected by sensory panels varies between fish species, and is around 10-15 mg N/100g.

In fresh water fishes that lack TMAO, ammonia is produced by deamination of the amino acids by microbes. The compounds produced due to reduction of TMAO such as TMA, DMA, MA and ammonia are volatile and produce off odour of spoiling fish. These compounds are collectively called as "volatile base nitrogen". TVBN is also used as an indicator of microbial spoilage of fish. TVBN is a better indicator of spoilage than TMA because it includes intermediate products and the end products. The level of TVB-N in freshly caught fish is between 5 and 20 mg N/100g and the limit of acceptability of ice stored fish is 30-35 mg N/100g.

During chilled or frozen storage of fish, when bacterial growth is inhibited, the reaction of conversion of TMAO into TMA is replaced by slow conversion of TMAO into DMA and formaldehyde. The formation of these products may cause severe quality changes such as increased denaturation, changes in texture and loss of water binding capacity or spoilage during prolonged storage.

ii. Urea Degradation

Urea is an osmo- regulatory substance present in the tissue of cartilaginous fishes like sharks, skates and rays. It is decomposed to carbon di oxide and ammonia by the microbial enzyme, urease. This process benefits the bacteria as they use the ammonia for respiration. The production of ammonia also raises the pH of the substrate which promotes the growth of many urea degrading bacteria inhibiting competition by many other species. The urea degradation is associated with bacteria such as *Helicobacter pyroli, Klebsiella pneumoniae, Proteus* spp. and *Micrococcus luteus*.

iii. Free Amino Acid Degradation

The free amino acids are present in the fish tissue in small quantities. After the death of fish, due to autolysis, free amino acids are formed from tissue proteins resulting in nutrient rich medium for microbial activity. They are also formed from hydrolysis of tissue proteins by proteolytic enzymes secreted by microbes such as *Pseudomonas, Sarcina* when the available free amino acids are scanty or exhausted. The free amino acids are absorbed into the microbial cell where they are metabolised through decarboxylation and deamination by specific enzyme system of microbes. The metabolic end products come out of the microbial cell. Degradation of free amino acids takes place in three ways:

a. Deamination

Deamination is the process of the removal of amino group from the amino acids and the amino group is then reduced to ammonia. The microorganisms associated are *Pseudomonas,* Enterobacteriaceae and Lactic acid bacteria. Deamination occurs in four ways.

Oxidative Deamination

It takes place in the presence of oxygen with production of keto acid and ammonia. The keto acid eventually turns into aldehydes, lower fatty acids and carbon di oxide.

Reductive Deamination

It occurs in the presence of hydrogen producing a saturated fatty acid and ammonia.

Hydrolytic Deamination

In this process, free amino acids are hydrolysed in the presence of water with the production of hydroxyl acid and ammonia. The hydroxyl acid is decarboxylated with theliberation of carbon di oxide and production of alcohols.The alcohol is oxidised further to aldehydes, ketones and lower fatty acids.

Desaturative Deamination

The amino acid is converted to unsaturated fatty acid with the formation of ammonia.

Table 2.1: Formation of Biogenic Amine from Amino Acids

Amino Acid Precursor	*Amino Acid Decarboxylase*	*Biogenic Amines*
Histidine	Histidine decarboxylase	Histamine
Tyrosine	Tyrosine decarboxylase	Tyramine
Tryptophan	Tryptophan decarboxylase	Tryptamine
Lysine	Lysine decarboxylase	Cadaverine
Phenylalanine	Phenylalanine decarboxylase	Phenylethylamine
Arginine	Arginine decarboxylase	Agmatine
Ornithine	Ornithine decarboxylase	Putrescine, Spermidine, Spermine

b. Decarboxylation

Decarboxylation is the removal of the carboxylic group (-COOH) from the amino acids with the production of carbon di oxide and amines. These amines are called biogenic amines and include histamine, cadaverine, putrescine, agmatine *etc.* These biogenic amines are biologically active low molecular weight nitrogenous compounds. The biogenic amines formed from amino acids and the respective enzymes are given in Table 2.1.

c. *Deamination and Decarboxylation*

Here, both the amino group (-NH_2) and carboxylic group (-COOH) are removed from the amino acids. It occurs in three ways *viz.* (i) oxidative deamination and decarboxylation, (ii) reductive deamination and decarboxylation and (iii) hydrolytic deamination and decarboxylation.

2.5.2 Formation of Volatile Sulphur Compounds

Volatile sulphur compounds are typical components of spoiling fish and most bacteria identified as specific spoilage bacteria produces one or other volatile sulphides. *S. putrefaciens* and some Vibrionaceae produce H_2S from sulphur containing amino acid, l- cysteine. *Alteromonas* and *Pseudomonas* are active sulphide producers. Methylmercaptan (CH_3SH) and dimethylsulphide ($(CH_3)_2S$) are formed from methionine (Table 2.2).The volatile sulphur compounds are foul smelling and can be detected at very minute levels.

Table 2.2: Typical Spoilage Compounds Formed during Spoilage of Fresh Fish (Stored aerobically or packed in ice or at ambient temperature)

Specific Spoilage Organism	*Typical Spoilage Compounds*
Shewanella putrefaciens	TMA, H_2S, CH_3SH, $(CH_3)_2S$, Hx
Photobacterium phosphoreum	TMA, Hx
Pseudomonas spp.	Ketones, aldehydes, esters, non- H_2S sulphides
Vibrionaceae	TMA, H_2S
Anaerobic spoilers	NH_3, acetic, butyric and propionic acid

2.5.3 Hydrolysis of Complex Tissue Components into Simpler Substances and their Utilisation

When the simple substances such as TMAO, urea and free amino acids are completely exhausted, the microbes are start to hydrolysing complex tissue components such as proteins and lipids to produce simpler substances. For utilising protein, proteolytic group of bacteria secrete proteolytic enzymes from their cells to the surroundings.These enzymes hydrolyse protein into amino acids. These amino acids enters the microbial cell and are metabolised and the end products are liberated out of cell.*Alcaligenes, Acinetobacter, Aeromonas, Bacillus, Corynebacterium, Flavobacterium, Lactococcus, Lactobacillus, Pseudomonas* and *S. putrefaciens, Sarcina* are associated with proteolysis in fish.

The lipolytic microbes secrete lipolytic enzymes to the surrounding and hydrolyse fats into fatty acids and glycerol. Subsequently, they are utilised by microbes. The accumulation of free fatty acids in fish tissue leads to the development of rancid flavour in fish. This rancidity is called "hydrolytic rancidity". However, when compared to protein hydrolysis, lipid hydrolysis is negligible. Many of the proteolytic bacteria are found to be lipoplytic. Organisms of genera *Pseudomonas, Alcaligenes, Staphylococcus, Serratia* and *Micrococcus* are lipolytic bacteria.

2.5.4 Breakdown of Nucleotides

Hypoxanthine which cause bitter off flavour in fish is formed by autolytic or microbial activity. Several spoilage bacteria such as *Psuedomons, Shewanella putrefaciens* and *Photobacterium phosphorium* produce hypoxanthine from inosine or inosine mono phosphate.

Table 2.3: Substrate and Off Odour/Off Flavour Compounds Produced by Bacteria during Spoilage of Fish

Substrate	*Degradation Products*
TMAO	TMA
Cysteine	H_2S
Methionine	CH_3SH, $(CH_3)_2S$
Carbohydrates and lactate	Acetate, CO_2, H_2O
Inosine, IMP	Hypoxanthine
Amino-acids (glycine, serine, leucine)	Esters, ketones, aldehydes
Amino-acids, urea	NH_3

2.6 Spoilages and Quality Indices in Frozen Fish

2.6.1 Freezing Process

The freezing process is undoubtedly one of the most important factors connected with the physical, chemical and microbiological qualities of frozen fishery products. Fish contains about 75-80 per cent by weight of water. The process of freezing converts most of the water into ice. Water in fish contains dissolved and colloidal substances which depress the freezing point below 0°C and the extent of depression of freezing point is proportional to the concentration of the solute. About 75-80 per cent of the water is frozen between -1°C and -5°C. This temperature range is known as the critical or freezing zone. During freezing, water is gradually converted to ice and the concentration of dissolved organic and inorganic salts increases, depressing the freezing point continuously. So even at a temperature of -25°C only 90-95 per cent of water is actually frozen. The bound water, which is the water chemically bound to specific sites such as carbonyl and amino groups of proteins, is not available for freezing. During freezing the temperature falls rapidly to below 0oC and then falls very slowly until most of the water is changed to ice and therefore the temperature drops quickly. Since spoilage continues at temperatures just below 0°C, it is important to pass the critical zone quickly. Fast freezing will control the extent of protein denaturation by limiting the exposure of tissues to the salts concentrated by ice formation. The longer the exposure, the larger is the damage to the proteins and to other enzyme systems and their substrate.

The size of ice crystals formed in frozen fish muscle also varies according to the rate of freezing. It is possible that the slow removal of heat, by slow freezing will produce ice crystals of large size and few in numbers which can cause rupture of cell walls resulting in fluid loss and textural changes on thawing. But rapid removal of heat by fast freezing produces large numbers of small crystals thus reducing the possibility of tissue shrinkage and rupture. Cell damage will cause increased thaw drip and loss of important constituents of fish like vitamins, minerals, amino acids and compounds associated with characteristic aroma and flavor of specific species. The time taken to lower the temperature of the thermal center to -20°C is normally termed as the freezing time.

State of Rigor at the Time of Freezing

The degree of biochemical changes during freezing is affected by the condition of the fish at the time of capture and the state of rigor at the time of freezing. The longer the rigor lasts, the better the keeping quality of the flesh. Frozen fish blocks made from post rigor fish are more desirable. Drip loss of fillets and blocks prepared from pre-rigor fish is much higher than in those prepared from post-rigor fish. Thus it is advantageous to extend the rigor mortis stage to just before freezing to enhance the freshness and keeping quality.

2.6.2 Frozen Storage

Once the fish is frozen, biological spoilage caused by the exogenous enzymes of the bacteria ceases. On freezing the autolytic changes caused by the breakdown of the biochemical constituents of the fish by the intrinsic enzymes tend to be relatively slow and contribute little to the loss of quality. The major changes in frozen seafoods fall into three major categories:

a. Denaturation of proteins,
b. Color and flavor changes
c. Breakdown of the connective tissue

i. Protein Denaturation in Freezing and Frozen Storage

Protein denaturation in frozen fish muscle is the cause of a spatial rearrangement of protein molecules, due to the formation of ice crystals within the muscle fibrils. This results in the molecular uncoiling followed by interlacing of various fractions among themselves and with other biochemical components of the tissue. Denaturation brings about definite changes in the chemical and physical properties of native proteins such as solubility, hydration, optical properties, digestibility *etc.* During freezing denaturation no chemical breakdown are known to take place and the functional groups of the amino acids are not disrupted. Hence, proteins do not seem to be affected in their nutritive properties. Prolonged icing before freezing can cause damage to the proteins mainly by the action of proteolytic enzyme and psychrophilic bacteria. This can be manifested later in the quality characteristics of the resulting frozen fish like color, flavor and texture.

Prolonged frozen storage leads to increased protein denaturation resulting in cell damage, loss of protein solubility and deduction in hydration capacity of proteins. Many secondary changes following the protein denaturation are caused by interaction of -SH groups, both of protein and non-protein origin. Aggregates of protein molecule can occur due to sulphydryl - disulphide exchange. Most of the fish protein denatured during frozen storage is soluble when disulphide bond reducing agent like beta mercapto ethanol is combined with sodium dodecyl sulphate (SDS)

Theories of protein denaturation in frozen fish emphasizes the following:

1. The redistribution of water and associated concentration of solutes in the tissue
2. Lipid hydrolysis and/or oxidation and interaction of products with proteins.
3. The formation of formaldehyde by TMAO demethylase.

Role of Cryoprotectants

Small molecules like polyphosphates, sugars, polyalcohols, dicarboxylic acids, *etc.* are known for their protective effect on fish muscle proteins. They act by association with protein via ionic or hydrogen bonding and thus increasing the electrostatic repulsion. The osmolytic cryoprotectants like polyols and sugars (*e.g.* sorbitol, sucrose) alter formation and growth of ice crystals or stabilize protein-water interaction. High molecular weight polymers like maltodextrins, alginates, and carrageenan and carboxymethyl cellulose increase molecular density and solvent viscosity and thus lower protein aggregation.

ii. Oxidation of Lipids in Frozen Fish

Oxidation of lipids is a major cause of off-flavors and textural changes of frozen seafood. The oxidation of unsaturated fatty acids involves the formation of free radicals and hydro peroxides. The study of lipid oxidation in fishery products emphasize the non-enzymatic or auto-oxidative processes. Non-enzymatic lipid oxidation is enhanced by metal ions like iron, copper and cobalt. In vivo, the bound forms of iron like haeme, non-haeme iron and chelation complexes like histidine-iron may play an important role in catalyzing lipid oxidation.

Lipid oxidation in frozen fish tends to occur at a more rapid rate in the dark muscle and skin and is stimulated by mincing the tissue. The rate of lipid oxidation is also directly related to the fat content. Free fatty acid formed by hydrolysis can contribute to the texture toughening by influencing protein denaturation. It can also enhance lipid oxidation leading to flavor deterioration. Protein and lipids can interact by hydrophobic interactions The free fatty acids are more prone to peroxidation than esterified fatty acids. Recent studies have indicated that enzyme catalyzed reaction may also contribute to the lipid oxidation during the early stages of storage by the enzyme lipoxygenase. This may explain why fish stored in ice for a few days prior to freezing become rancid faster during frozen storage.

iii. Colour Changes

Characteristic colour in seafood is an important factor for consumer acceptance. Loss of the pink color in crustacean shellfish results from the changes in the carotenoid pigments:

Beta carotene (red) → astaxanthin (pink) → astacene (orange-yellow)

Yellow discoloration of lipid rich regions in frozen fish also results from lipid oxidation.

Blue/Black and Brown Discolouration in Frozen Lobster Tails

As in prawns, black spot development occurs in lobster tails also. This can be prevented by keeping the lobster's alive upto the time of processing. Brown discolouration develops during frozen storage at the cut end of the meat portions probably by reaction of the enzymes. Proper glazing, wrapping and low temperature storage are remedial measures.

Discolouration in Squid and Cuttle Fish

The important problem in frozen squid and cuttle fish is the yellow discolouration of the tubes and fillets. Bleeding immediately after catch and removal of appendages, inksac and gut contents followed by washing prevent yellowing. The semi-dressed material has to be stored in ice and water. Treatment with a solution of salt and citric acid is found to improve colour and texture of the squid and cuttle fish.

iv. Thaw Drip

The degree of protein denaturation is reflected in the amount of thaw drip exuded from the cells on defrosting the fish. Excessive drip can be related to (a) slow freezing, (b) undue temperature fluctuation during storage and (c) prolonged freezer storage. Drip loss results in a dry texture with loss of compounds contributing to flavor like amino acids, nucleotides, sugars *etc.* Variations in frozen storage behaviour depend on the cell fragility of the fish and the extent of biochemical changes that have taken place before freezing. Hypoxanthine and other nucleotides as well as trimethylamine are used as indices of freshness and pre-freezing condition as they do not seem to change during frozen storage. It is also important to prevent periods of thawing and refreezing. During thawing various enzymes diffuse out by membrane damage. The deteriorative reactions by these enzymes will continue even after refreezing. Partial or complete thawing and refreezing also increases the amount of drip.

v. Freezer Burn/Dehydration

Low humidity in the freezer along with fluctuation in temperature during long storage periods are causes for freezer burn. Under this condition, moisture is rapidly lost from the cells by sublimation. Thus frozen fish develops a chalky-white, dry and wrinkled appearance on the surface and further to yellow or brown discoloration. Weight loss by dehydration is proportional to the surface area and proper glazing and packaging can reduce this problem. Weight loss by dehydration should not be more than $50g/m^2/24hrs$.

Dehydration in Frozen Products

Dehydration is evident as white patches on the frozen fish. Proper packaging with exclusion of air pockets, sufficient glazing, maintenance of constant storage temperature and relative humidity are important to reduce dehydration. The maximum dehydration allowed in shrimp for export from India is 20 per cent by count.

vi. Black Spots in Shell - on Shrimp

Black spot formation or "melanosis" is a major problem in prawn freezing industry. This is an enzymatic reaction and requires access to oxygen in addition to the presence of heavy metals like copper and iron. Melanin pigments are produced by an oxidative reaction by tyrosinase on' tyrosine. By this discolouration the eating quality is not lost, but it masks the appearance. These prawns can be packed as peeled meat. But this will diminish the unit value. One method of prevention of melanosis is cutting off access to oxygen. This can be done by keeping the material in ice and water with a layer of water above the material. As the enzyme is more concentrated on the head portion, removal of head immediately after catch and subsequent icing will prevent black spot formation. *M. affinis* is most susceptible to black spot formation followed by *P. indicus, M. dobsoni* and *P. stylifera*. Treatment with potassium metabisulphite (0.2 - 0.5 per cent for 1 to 2 minutes) is beneficial to reduce black spot formation. But higher levels of sulphite cause bleaching of the shell colour. By treatment, the level of sulphite as SO_2 should not exceed 100 ppm in raw meat and 30 ppm in cooked meat. The maximum black spot permitted is 10 per cent by count in shell -on types and 5 per cent by count in peeled type.

vii. Weight Loss during Thawing

On thawing, frozen seafoods lose some weight as thaw drip consisting of water containing soluble nutrients and flavour bearing components. The weight loss due to drip is 5 per cent in headless, 10-15 per cent in peeled and deveined and 7 to 10 per cent in cooked shrimp. It also depends on size: bigger the size, smaller the loss. It is also dependent on species (*M. dobsoni* and *P. stylifera* recorded highest). It increases with pre-freezing ice storage period. Drip loss is prevented by treatment with polyphosphate. The maximum residual phosphate permitted is 5gm/kg as P.

viii. Bacterial Contamination

In addition to good organoleptic characteristics, frozen prawns/fish have to be free from excessive numbers of bacteria. Some types of bacteria, for *e.g.* coliforms, *E.coli, Staphylococcus etc.* shall be present only in very limited numbers and some types for *e.g. Salmonella, V. cholerae, Listeria etc.* shall be totally absent. Precautionary sanitary practices like chlorination of water supplies both for use in processing and ice manufacture, application of regular cleaning schedules, strict enforcement of workers' hygiene *etc.* They are very important to keep down bacterial contamination.

ix. Foreign Materials (filth)

Materials like flies, cockroaches (whole or body parts), hairs, fibre pieces, bits of paper and excessive sand can often be encountered in frozen shrimp depending upon hygienic conditions. These, in general, are termed as filth. There has been some rejection of Indian shrimp in USA due to filth. Careful washing and packing of the raw material and exclusion of flies, rodents *etc.* In the processing premises will solve the problem of filth to a large extent. The avoidance of peeling on floor and use of potable water for washing and proper washing of whole prawns with good quality water can reduce sand content.

x. Contamination with Heavy Metals

Heavy metals have been recognised as serious pollutants of the aquatic environment. They include metals like chromium, mercury, cadmium, cobalt, nickel, copper, zinc, lead and tin and metalloids like arsenic and selenium. The levels of these metals are within the limits in fishes. But cadmium has been reported to be above the limit in samples of whole squid/cuttle fish. Studies have shown that cadmium is mainly concentrated in the liver part and this problem can be avoided by careful removal of liver.

xi. Decomposition and Indole in Shrimp

Decomposition is another serious problem in seafoods. Seafood spoils very quickly if proper care is not taken during its processing, storage and transportation. The maximum decomposition permitted in frozen shrimp is 5 per cent by count. Decomposition is judged by various chemical quality indices such as indole, TVN, hypoxanthine *etc.* Some of the importing countries have fixed tolerance levels for indole. The maximum permitted limit of indole in shrimp is 25 microgram/100gm.

xii. Histamine in Fish

In scombroid fishes like tuna, mackerel *etc.*, histamine is formed during spoilage. Histamine is formed by bacterial action on the amino acid, histidine present in scombroid fishes. Above certain levels, histamine causes food poisoning. The maximum permitted level of histamine in fish is 20mg/100gm.

xiii. Pesticide Residues

In recent years, aquaculture is given greater importance in our country. As most of the aquaculture is carried out either in paddy fields or adjacent areas, the problem of pesticide accumulation in these cultured fishes cannot be ruled out. Some of the most important groups of chemicals widely used as insecticides are DDT and its derivatives (DDE, DDD, or TDE), aldrin, dieldrin, benzene hexa chloride (BHC) and polychlorinated biphenyl (PCB). Most of the countries which import seafoods have specified the maximum permissible limits of certain pesticides in the seafoods. The limits prescribed by USFDA are 0.3 ppm for Dieldrin and Endrin, 5.0 ppm for DDT, DDE and TDE and 0.3 ppm for Heptachlor.

xiv. Poor Quality Packaging Materials

The packaging material used for packing processed sea foods shall be able to with stand the stress and strain of transportation and storage under cold conditions.

2.7 Spoilages and Quality Indices in Canned and Retort Pouch Processed Fish

2.7.1 Bacterial Problems in Canned Products

All canned products are to be commercially sterile. Commercial sterility means that thought some bacteria may be present in the can, they may not be able to multiply and spoil the contents of the can under the environmental conditions available in the can. Sterility (Commercial or absolute) is tested after incubation of the cans at 37°C or 56°C for 14 or 4 days, and examining for growth in nutrient media. As a whole, 3 per cent of canned shrimps are shown to be not commercially sterile. Bacterial problem may be due to insufficient sterilization and post process contamination from cooling water. It is essential to see that the processing time and temperature are kept up correctly. The processing line needs to be kept in clean conditions and there should not be any hold up. Post process contamination usually occurs from cooling water. Cooling water shall be of sound in bacterial quality and shall contain free available chlorine. Contamination can take place through faulty seams as also through normal seams. Seams, though normal, can always allow sucking in of water at the time of cooling because of the strain due to temperature and pressure differences. The preventive step for faulty seam is checking of seam after every 200 cans.

Classification of Spoiled Cans

Flipper: This is a can of normal appearance in which one end flips out when the can is stuck against a solid object. The end snaps back to the normal position when very light pressure is applied

Springer: This refers to can in which one end is bulged but can be forced back in to normal position where upon the opposite end bulges

Soft Swell: It is a case in which the bulged ends can be moved by thump pressure but cannot be forced back to the normal position

Hard Swell: This is one in which the ends of the can are permanently and firmly distended

The chief defects and causes of spoilage may be listed as follows:

1. Microbial spoilage: This may result due to:

   Under processing

   Inadequate cooling

Leaker infection/Leakage through seams

Pre-process spoilage

2. Chemical spoilage

   Internal corrosion giving rise to hydrogen swells or pin holing

3. Physical: causes due to

   Faulty retort operation

   Under exhausting

   Over filling

   Internal vacuum too high (Panelling)

   Use of cans of inadequate substances

   Rough handling

4. Miscellaneous like: Rust, Damage *etc.*

a. Under Processing

Any pack, which suffers spoilage as a result of the activities of microorganisms surviving the heating process, can be termed under processed. In under processed packs, the activity of the surviving microorganisms may result in gas production, which causes the can to become a 'swell' (which is due to the growth of *Clostridium botulium*) or the contents may undergo acidification or some other undesirable changes affecting quality, but gas is not produced. When growth of microorganisms occurs without gas production, the affected cans have a normal appearance externally and spoilage is only detectable after can has been opened. Such spoilage is called as flat sour spoilage, which is due to the growth of *Bacillus stearothermophilus.*

b. Inadequate Cooling

Some bacteria such as thermophiles multiply rapidly in a high range of temperature and failure to cool cans immediately after processing to a temperature of about 35°C may lead to serious spoilage.

c. Leakage through Seams

The microorganisms infecting canned fish as the result of post - process leakage of the containers may be of widely varying types. The main source of the organisms is the cooling water. In ideal seaming, the edges of the cover and the body are properly folded to from five folds of metal. In improper seaming, the cover hook is compressed against the folded body hook creating an opening between the body and the cover for a portion of the perimeter through which spoilage microorganisms enter the can.

d. Pre-process Spoilage

Spoilage of this type is an example of faulty canning practice, whereby bacterial development in the fish is permitted during the preparation period. Processing may subsequently sterilize the pack, but the liberation of gas produced by the organisms during the lag period before processing may cause swelling or flipping of the cans.

2.8 Defects in Canned Fishery Products

Common spoilage in canned products are blue/black discolouration in canned shrimp and crab meat, underweight in canned shrimps, excessive water content in oil packs and high sand content in canned mussels and clams.

a. Blue/Black Discolouration in Canned Shrimp and Crab Meat

Discolouration occurs when the concentration of copper and/or iron exceeds certain critical levels. The muscle takes up copper and iron during processing from contact surfaces, water, ice, salt, citric acid, *etc.* and they combine with sulphur present in shrimp to form sulphide. Due to sulphur resistant lacquer, any imperfection will expose iron/tin layer of the container. The sulphur containing components released during thermal processing will react with iron and forms iron sulphide, which is black in colour. This reaction takes place rapidly if alkaline conditions have developed in the raw material due to spoilage. Discolouration due to copper and/or iron sulphides can be prevented by adjusting the titrable acidity in the fill brine and also by addition of EDTA disodium salt in the fill brine at the rate of 50 mg per cent. Very fresh shrimps require only lower quantities of citric acid to adjust the titrable acidity to 0.06 per cent whereas raw material stored in ice requires more acid. Crabs too contain higher levels of copper. Blueing can be prevented by either proper bleeding at the time of butchering or by addition of EDTA salt.

b. Green Discoloration

It occurs in cooked tuna. The TMA content in cooked tuna is closely related to the degree of green colour and that the development of green colour is closely correlated with the TMAO content of raw fish. TMAO content of more than 13mg per cent N was found to cause greening, a level below this occurrence of greening is unpredictable. It is recommended that tuna with high concentration of TMAO and myoglobin shall not be used for canning.

c. Brown Discolouration

Brown discolouration is caused by reaction of protein or amino acid with product of lipid oxidation. It is observed in variety of processed products including white pomfret, sardine *etc.* Discolouration due to protein-lipid browning is greater in fatty fish than lean fish.

d. Underweight in Canned Shrimps

Canned shrimps should have moisture content of 72 per cent, which is called as equilibrium moisture level. To get correct drained weight, shrimps should be blanched to the equilibrium moisture content. If the moisture level is higher and the material is underblanched, more water is removed from the meat during processing leading to the underweight of the canned product. Hence, it is always essential to standardize blanching conditions and to follow it strictly.Particular attention is required to keep up the concentration of salt and citric acid in the blanching brine, as also the time of blanching to obtain correct drained weights.

e. Excessive Water Content in Oil Packs

Fish canned in oil shall not contain more than 10 per cent water in the drained liquor. More water in the fill medium is a defective quality factor as it amounts to cheating the consumer and lowering the shelf life considerably by corroding the can and causing deterioration of the contents. Proper precooking and draining of the cook drip prevent the excess water content in the pack.

f. High Sand Content in Canned Mussels and Clams

It is a serious defect in canned clams and mussels. Removal of sand from the raw meat during processing is particularly difficult. The live material shall be subjected to purification in bacteriologically clean water for 24–48 h., and so that there is an appreciable reduction in the bacterial load sand content of the meat.

Fish canned in oil shall not contain more than 10 per cent water in the drained liquor. More water will lower the shelf-life considerably by corroding the can and causing deterioration of the contents. This defect is due to insufficient pre-cooking and can be prevented by proper precooking and draining of the cook drip.

g. Struvite Formation

The formation of magnesium ammonium phosphate hexahydrate (struvite) crystals is frequently found in canned tuna. Although these crystals have no taste or odour and are otherwise quite harmless, they are often mistaken for glass fragments by the consumer. Magnesium (usually from water used) combines with ammonia generated from the fish muscle to form struvite. This resultant compound gradually crystallises. The pH is important, the phosphates being soluble in small amounts below pH 6.5 and in larger amounts below pH 4.

h. Honey-combing

Honey -combing is often encountered in canned tuna. It consists of pits in the fish tissue, usually between the meat layers and is seen after pre-cooking. Sometimes the pits are also seen on cut surfaces. A cross section of extensively pitted tissue has an appearance of an empty honeycomb. It is always considered as a sign of advanced decomposition.

i. Stack Burning

Stack burning is caused by over processing. A considerable amount of heat is retained over a long period when canned products are stacked

j. Perforation and Corrosion

Perforation and Corrosion in can is prevented by air should be expelled from product and headspace as completely as possible. It is dependent on the temperature. Cans should be thoroughly cooled before packing or stacking to r educe corrosion.

k. Gaseous Spoilage

Can appears like Swelled or "bulging. It is caused by spore-formers of the anaerobic or facultative anaerobic types. Organisms found quite commonly are *Clostridium welchii* and *Clostridium sporogens.* and gas-forming heat-resistant organism is *Clostridium botulinum.*

l. Non-Gaseous Spoilage

There is no external indication of non-gaseous spoilage. It is caused by aerobic spore formers, *Bacillus cereus, B. mesentericus* and *B. vulgatus.* Storage at temperatures between 4° and 3°F. will greatly reduce the possibility of flat souring.

m. Mush

It is Flabby condition met with some species of pilchards caught at the end of its spawning. This is caused by the invansion of prasitic protozoan *chloromyxum* whish decomposes the fish meat during storage such that it becomes entirely soft during canning.

n. Curd and Adhesion

'Curd' is precipitated protein often found in canned mackerel and salmon. This is more common with salmon, which is generally canned without pre-cooking. The meat coagulated by heat adheres to the inner side of the can ends and presents a poor appearance on opening the can. The lacquer may get peeled off while removing the curd from the can ends. Use of raw fish, which is not very fresh, and, inadequate brining and pre-cooking are some of the reasons responsible for formation of curd. It can be prevented if the raw fish is soaked in 10-15 per cent brine for 20-30 minutes followed by thorough washing before filling.

2.9 Spoilages and Quality Indices in Cured Fish (Salted, Dried and Smoked)

a. Pink Spoilage

A common type of spoilage in cured fish usually encountered in tropical countries like India is the formation of pink slimy patches on their surfaces emanating unpleasant odour in advanced stages. This is caused by several species of coloured halophilic bacteria present in the salt used for curing. Pink/red spoilage is mainly due to the presence of halophilic bacteria (*Halobacterium salimaria, H. cultrirubium, Sarcina morrbuae* and *S. litoralis* from the salt. These bacteria are not harmful by nature. They are aerobic and proteolytic in nature, grows best at 42°C and at 10 per cent salt concentration by decomposing protein and giving out an ammoniacal odour. Use of good quality salt will avoid this defect to a great extent. This spoilage is usually found in heavily salted fish and is absent in unsalted fish.

b. Dun

Dun is another type of spoilage which manifests itself as chocolate brown spots caused by the growth of a halophilic mould *Sporendoneina epizoum* giving an unpleasant appearance. Fungus usually grows well on unsalted and salted dried fish, which has high moisture, content. Moulds usually grow at relative humidity above 75 per cent. The optimum temperature for growth is 30 –35°C. In salted fish, brownish black or yellow brown spots are seen on the fleshy parts. This gives the fish a very bad appearance.

The climatic conditions prevailing in tropical countries are favourable for the growth of this mould. Dun can the prevented by dusting the dried product at the time of packaging with a mixture of 3 parts of sodium propionate and 97 parts of sodium chloride at the rate of 10 per cent by weight.

c. Rancidity

This is caused by the oxidation of fat. Rancidity is more pronounced in oil rich fishes like mackerel, sardine *etc.* The unsaturated fat in the fish reacts with the oxygen in the atmosphere forming peroxides, which are further broken down into simple and odoriferous compounds like aldehydes, ketones and hydroxy acids, which impart the characteristic odours. At this stage the colour of the fish changes to brown. This is known as rust. This change results in an unpleasant flavour and odour to the product, leading to consumer rejection. Fatty fishes continue to become more and more rancid during storage. Certain impurities in salt and traces of copper accelerate this. Rancidity reduces the nutritive value and consumer acceptance and is a serious problem in cured fatty fishes.

d. Insect Infestation

Spoilage due to insect infestation occurs

1. During initial drying stages
2. During storage of the dried samples.

The flies, which attack the fish during the initial drying stage, are mainly blowflies belonging to the family Calliphoridae and Sarcophagidae. *Chrysomya megacephala* is the major species, which causes damage. These flies are attracted by the smell of decaying matter and odours emitted from the deteriorating fishes. During the glut season when the fish is in plenty and some are left to rot, these flies come and lay their eggs. The eggs develop into maggots, which bury within the gill region and sand for protection from extreme heat. They develop mainly when conditions are favorable with adequate moisture and intermittent rain. This results in loss of nutritive value of the fish and economic losses to the processor.

The most commonly found pests during storage are beetles belonging to the family Dermestidae. Beetles attack when the moisture content is low, especially when the storage time is for long. The commonly found beetles are *Dermestes ater, D. frischii, D. maculates, D. carnivorous* and *Necrobia rufipes*. The larva causes most of the damage by consuming dried flesh until the bones only remain.

Mites are important pests, found infesting dried and smoked products. They are very minute and bring about powdering of the product thereby giving it a white appearance. *Lardoglyphus konoi* is the commonly found mite in fish products.

Infestation can be reduced by:

1. Proper hygiene and sanitation
2. Disposal of wastes and decaying matter
3. Use of physical barriers like screens, covers for curing tanks etc.
4. Use of heat (45°C) to physically drive away the insects and kill them.

e. Fragmentation

Denaturation and excess drying of fish results in breaking down of the fish during handling. Fish can become brittle and is liable to physical damage when handled roughly. It is necessary that fresh fish is to be used as raw material to ensure a good quality finished product.

2.10 Quality Control and Safety Aspects of Cured Fish

The cured fish products are easily prone to spoilage and deterioration if they are not processed and stored well. The main problems affecting cured fish are, histamine, mycotoxins, 3,4 benzopyrene, spoilage due to oxidation, and halophilic bacteria and mould.

Histamine is a biogenic amine. It is an index of bacterial spoilage as well as a risk factor to the health of consumers. Maximum histamine of 20 mg/100 g fish has been fixed as the rejection level. It is mainly seen in scomberoid fishes. Cured fish are easily prone to moulds and fungal attack. The product must be dried fully and must not be left uncovered in damp places. Lightly infected fish can be cleaned with brine solution and redried or smoked again. The main fungi isolated from dry fish were *Aspergillus, Mucor* and *Pencillium* species. Some of the species are pathogenic in nature and known to cause food spoilage and hence of significance in food safety. Care should be taken to see that highly infected fish is discarded because of the risk of mycotoxins, which are harmful to the consumers. Aflatoxin is a mycotoxin produced by the fungus *A. flavus*. It is a potential carcinogen.

Poly Aromatic Hydrocarbons (PAH) contain many highly carcinogenic substances. 3,4 benzopyrene is a carcinogen that has been detected in different smoked products. Nitroxo (Nox) substances in smoke imparts a pink color to the flesh of smoked fish. Nox substances are also capable of forming carcinogenic N-nitrosamines by reactions with the amines in the fish.

2.11 Spoilages and Quality Indices in Surimi

2.11.1 Quality Assurance in Surimi Processing

The Total Quality Management programme consists of process control and product evaluation. The quality inspection programme will take the form of an HACCP system which would identify critical control points along the processing line and establish operating parameters.

Process Controls

1. The temperature, pH and weight of the whole round fish are recorded at the time of delivery. The fish is held as cold as possible without freezing until processing begins.
2. Dressed fish should be inspected for proper removal of viscera, backbone and belly flaps when they leave the heading and gutting machine.
3. Proper functioning of the deboner is very important in the operation of a surimi line.
4. In the leaching tanks, where the mince is mixed with water, the water to mince ratio must be maintained at the optimum. Water/mince ratio may vary from 2:1 to 10:1 depending upon the species of fish, freshness *etc.* Temperature of water should not exceed 5°C.
5. In the refining stage, control the speed of refiner is very critical to check heat gain and for proper separation of washed mince.
6. During dewatering the water content should be checked at periodic intervals by properly adjusting the dehydrator.
7. Proper ratio of cryoprotectants, time of mixing and temperature are to be taken care of during the mixing of ingredients. A mixture of sugar (2-4 per cent), sorbitol (4 per cent) and polyphosphate (0.1 per cent - 0.3 per cent) is usually employed to prevent protein denaturation during frozen storage.

2.11.2 Quality Evaluation of Surimi

Colour

Colour is measured using a whiteness meter.

Water Content

It is measured with an infra red moisture meter.

Gel Strength

Partially thawed surimi is mixed with 2-3 per cent salt and then filled in a PVC casing or stainless steel tube of 30 mm dia and 20 cm length. The tubes are sealed at both ends and heated at 90°C for 30 minutes, cooled and stored at 20°C until testing.

Specimens of 25 mm length are cut out from the cylindrical sample. The sample is loaded in a texturometer (or rheometer) and then subjected to compression using a probe having hemispherical (5mm dia) edge, moving at regular speed. Distance traveled by the probe (deformation) multiplied by the force at rupture of the sample gives the gel strength.

Folding Test

Slices of 3 mm thickness are cut off from the cooked sample. They are folded into quadrants. Grading is done according to the extent of crack developed on folding.

AA – No crack after folding twice

A. – No crack after first folding but crack occurs on second folding

B. – Crack occurs gradually on first folding

C. – Crack occurs immediately on first folding

D. – Breaks by finger press even before folding

Sensory Score Test (Bite Test)

Extremely firm	:	10
Very firm	:	8
Moderately firm	:	6
Slightly soft	:	4
Very soft	:	2
Extremely soft	:	1

Impurities Test

Scoring is done by counting the number of black skin and bone residues in a 10cm x 10cm field, which is prepared by spreading 10 g sample in 1 mm thickness. Particles over 2 mm in length are counted as (1) and particles under 2 mm in length are counted as 1/3.

Proximate Composition

Protein, fat, moisture and ash are estimated by standard methods.

Properties of Surimi

The surimi process involves washing out the blood, pigments, sarcoplasmic proteins and odour bearing compounds. The washed mince, which has a high concentration of myofibrillar proteins, readily form a gel as a result of unfolding and cross linking of actomyosin, the major muscle protein complex. Gel formation occurs when surimi is heated at 80-90°C but also takes place slowly at 40-50°Cand even at 0°C when held for a long time. Surimi that has been initially set at 40-50°C, followed by heating at 80-90°C gives a stronger gel. However, when the gel is heated to 60-70°C the extent of cross-linking is reduced, leading to softening of the gel (Modori phenomenon).

2.12 Quality Defects in Coated Fishery Products

Coated product is one, which is coated with another food stuff seafood specialities, fish portions (raw and precooked), shrimp, fish fingers, scallops, fish balls, fillets *etc.*, are the principal seafood products which are battered and/or breaded.

i. Shelling

1. Separation of breading from substrate due to uncontrolled release of moisture
2. Corrective action – Controlled release of moisture from substrate and breading system.

ii. Blow off

1. Breading blows off in the fryer
2. Corrective action – Monitor free water/ice on the surface of the substrate.

iii. Poor Adhesion

1. Breading system does not adhere to the substrate
2. Corrective action – Increase protein to perform binding.

iv. Gummy Interface

1. Area between surface of the substrate and breading system is gummy
2. Corrective action – Ensure product is fully cooked

Chapter 3

Quality Assessment of Fish and Fishery Products

3.1 Methods of Quality Assessment

The degree of spoilage or quality can be assessed in four different ways, namely;

1. Organoleptic assessment
2. Chemical/biochemical method
3. Instrumental method and
4. Microbiological method

3.2 Organoleptic/Sensory–Subjective and Objective

The quality and wholesomeness of food products are the primary concern of the consumers. Many methods are available to quantify the quality. The important methods are physical, chemical, microbiological and sensory. The first three methods require sophisticated equipment, laborious and time consuming experimental procedures and trained and skilled persons to carryout experiments. The results obtained need not actually depict the real condition of the product and it will not have sometimes any connection with the actual perception by the consumer. The actual consumer appeal can be reflected in sensory analysis.

Sensory analysis is concerned with measuring physical properties by psychological techniques. Sensory methods are used for measuring the properties that cannot be evaluated directly by physical or chemical tests. Conducting sensory tests may not seem difficult to the layman but it is not as easy as it seems. If proper attention is not given in sensory testing the data obtained will not reveal the true situation. It is necessary to be familiar with the techniques available, to know when and how to use them and to have a panel that has been carefully screened and trained.

The characteristics of food materials are evaluated depending on the stimulus received when one is testing the material, which are measured and quantified. A stimulus may be defined as any chemical or physical activator, which causes a response in a receptor. For example, eye is the receptor for light stimuli; nose for odour, tongue for taste *etc.* Six classes of stimuli exist. They are mechanical, thermal, photic, acoustic, chemical and electrical. An effective stimulus produces a sensation, the dimensions of which are quality, intensity, extension, duration, like and dislike. The sensation is measured by psychological procedures.

The least energy capable of producing a sensation is called absolute threshold. The least stimulus changes perceptible is termed difference threshold. For the study of food materials the senses of sight, hearing, smell, and taste and the four cutaneous (skin) senses *viz.* cold, warmth, pressure and pain are used.The senses of sight and hearing play a major role in regulating the function of human world. These senses help him in communication, navigation and detection of objects at a distance. Anyhow, man derives comfort from touch and pleasures of odour and taste. It is generally accepted that the stimulation of one sense organ influences to some degree the sensitivity of organs of another sense. Whether the influence is exerted upon the receptors or upon their central areas in the cortex is not known with certainty.

There are many methods of sensory tests. Laboratory taste panels are used to determine detectable difference and to measure the amount of certain properties, which cannot be measured readily by physical equipments. Consumer surveys are used to determine preference or acceptance to a particular product by the consumers. Neither of these methods can be used to do the task of the other. When a new product is tested by a laboratory panel the assumption that need be made is that the laboratory panel is judging the same basic variables with at least the same sensitivity as the ultimate consumer. Field-testing with untrained judges usually produces large variations in results and thus poor reliability.

a. Appearance

The quality factors associated with appearance influence food appreciation and preference by the consumer. Often colour changes are accompanied by desirable/ undesirable changes in taste, odour or texture. Consumer expect food to have certain colour, a deviation from the colour may introduce sales resistance. Food selection will be extremely difficult if colour discrimination is removed retaining other appearance factors. It has been noticed that food colour and colour of the environment in which the food is seen can increase or decrease our desire and appetite for it. Liking and disliking of a food is conditioned by its colour. We learn from our experience to associate certain colours as natural and predict rather precisely what properties a food will have from our memory of similar materials. Our reaction to unusual properties can be changed if they are explained to us.

Colour and appearance factors are important in the selection and acceptance of fresh fish and processed fish products. Changes in the appearance of the skin and the colour of the gills are closely associated with deterioration in quality. The appearance of the skin of fresh fish is bright and shinning with no bleaching which changes to waxy with slight bleaching and the fish which is not acceptable for human consumption has a dull, gritty appearance with marked bleaching and shrinkage. Various colour changes can be noticed in the gills during progressive spoilage. It changes from bright red, mucus and translucent to pink, mucus and slightly opaque during the early stages and then to gray, bleached, mucus opaque and thick and finally to brown, bleached, mucus yellowish gray and clotted when the fish is unfit for human consumption. The appearance of outer slime and eyes also can indicate the extent of spoilage in fish.

In general we recognize, discriminate and select food with eyes. Through conditioning and association we expect an item to have certain shape and colour, to have a specific odour, taste and texture. Appearance of the product is identified with previously experienced quality and serves as an indicator of good or bad, according to the products intended use.

b. Flavour

Flavour is the sensation realized when the food is placed in the oral cavity. It is dependent upon the reactions of the taste and olfactory receptors to the chemical stimulus.

c. Taste

The word taste means the sensory response to soluble materials in the mouth. This includes esthetic appearance also. The gustatory cells located in the taste bud are the actual receptors of the sensation. Taste buds have a limited life time and being renewed about every 210 hour. The response pattern appears to be stable. The anterior portion of the tongue is more sensitive to sweet stimuli, the posterior to bitter substances and the lateral portion to salty and sour stimuli. However there is considerable over lap.

Taste is a chemical sense, which is functioned as a warning and as a feeding mechanism. Soluble substances only can produce taste. Salivary glands are important in tasting particularly in dissolving tasteful substances and carrying them to the receptors. The high potassium content in saliva sensitizes taste receptors.

The four basic tastes are sweet, sour, bitter and saline. The numbers of distinct tastes are very large. It is believed that they are combinations of four basic tastes. The four fundamental taste qualities give variable sensations of pleasantness and unpleasantness depending on the concentration. The pleasantness of sucrose increased as the concentration increased and at a rather high concentration the pleasantness decreased slightly. If one taste is at or near the threshold and the other very strong, even the most sensitive subject will not perceive the lesser taste. Likewise, we reduce the strong sensation of one taste with another *e.g.* sugar in tea, sugar in lemonade *etc.* Although one taste may modify another, it does not neutralize it.

i. Sour

Sour is considered a true taste. Not all acids are sour; amino acids are often sweet and picric acid very bitter. Sourness is caused by the hydrogen ions. This does not entirely account for sour taste because at the same pH acetic acid is more sour than hydrochloric acid. Aliphatic carboxylic acids with long chain are sourer than those with short chain.

ii. Salty Taste

Inorganic salts produce saltiness but pure salty taste is that of sodium chloride. Potassium and lithium salts are salty but usually give a mixed taste. Potassium chloride is salty and bitter. Some lead and beryllium salts are sweet *e.g.* lead acetate and beryllium chloride. The particular taste depends not only on the salt employed but also on its concentration. The saltiness of different salt is additive.

iii. Sweet Taste

The sweet taste is produced by a variety of non-ionized hydroxy compounds *e.g.* alcohols, glycols, sugars *etc.* Certain electrolytes *e.g.* lead acetate and beryllium chloride are sweet. Sweet tasting compounds include sucrose, dulcin, saccharin, glycine, chloroform *etc.*

iv. Bitter Taste

The typical bitter stimuli are for alkaloids such as quinine, caffeine and strychnine. The bitter taste is often associated with compounds harmful to man. No single class of chemical compounds is characteristically bitter. Several electrolytes *e.g.* magnesium and ammonium compounds are bitter. Nitro compounds such as picric acid are bitter and the bitterness increases as the number of nitro group increases. Bitterness is a taste sensation, which can be evoked by the lowest concentration.

v. Interaction of Taste

Threshold tests for individual tastes are of interest, but we seldom encounter such problems in practice. Foods contain two or three or probably all basic tastes. A systematic study of the effect of basic tastes on each other was made and found that the presence of increasing amount of one taste raised the threshold of another taste. But in some cases this is not true. Example at low sugar levels the salt sensitivity is increased.

vi. Factors that Affect the Sensitivity of an Individual to Taste

There are many factors that affect the sensitivity of an individual to taste, some of which are discussed below.

(a) Effect of Disease

Disease and accident may result in the loss or decrease or alteration of sensations. This may be temporary or permanent. Irradiating the side of the tongue of a patient with X-rays reduced taste sensitivity to all tastes except sour. Recovery took about two months. In case of diabetics it is reported that sweet taste may be experienced in the absence of stimuli on the tongue. A bitter taste was reported in the case of jaundice. Intravenous injection of nicotinic acid can cause a giggle and metallic taste at the tip of the tongue. Drugs used to treat some metallic poisonings, such as pencillamine, can completely eliminate the sense of taste for few days.

(b) Effect of Sleep and Hunger

Lack of sleep does not affect the thresholds of salt and sweet but raise the sour threshold significantly. The sensitivity to the four basic tastes is greater at 11 - 11.30 am. There is significant decrease in sensitivity after meal. Loss of body water causes a decrease in sensitivity to salt but do not affect the sour threshold.

(c) Age and Sex

The newborn baby has little taste differentiation until about 35-40 days. A much higher sweet threshold has been found in 52-85 year old group. The decrease in tastes start in the late 50's and affect sour less than the other tastes. It has been found that woman have a higher sensitivity than men for sweet and salty but less for sour and no difference between the sexes for bitterness.

(d) Smoking

There is no difference between smokers and non-smokers in their thresholds for sweet, sour or salty but the mean threshold for bitter is significantly higher for smokers.

d. Odour

Odour is defined as that which can be smelled. Odour is the property of a substance that is perceived by inhalation in the nasal or oral cavity. Odour stimuli affect only a small area of the receptor cells located in the ceiling of the inner nose.

Odourless would be the verdict when such a sensory stimulus is not received. One problem is that odour cannot be measured quantitatively by nose. Subjective terminology such as camphor like, flowery, fruity *etc.* are commonly used to describe olfactory sensations. Some of the odour factors are fragrant (methyl salicylate) burnt (guaiacol), goaty or sulfurous (ethyl disulphide), etherish (1 propanol) sweet (vanillin), rancid (butyric acid), oily (heptanol), metallic (hexanol) and spicy (benzaldyhyde).

Olfactory adaptation takes place so rapidly that it can progress to a state in which the subject is completely insensitive to an odour. Some compounds, such as formaldehyde, diminish olfactory perception. Sensitivity to odour and degree of liking for odour differs markedly with age.

e. Texture

Texture deals with the sense of feel. Any sensation that may affect the skin or muscle endings may be considered with in the framework of texture. Practically the subject may be limited to hand specifically finger feel and mouth feel. It is suggested that such attributes of mouth feel as chewiness, fibrousness, grittiness, mealiness, stickiness, oiliness are essentially sensed by the muscular force applied in the process of mastication and are considered part of texture. The texture terms used above can be classified into three categories; mechanical characteristics, geometrical characteristics and other characteristics referring mainly to moisture and fat content

Objective measurements of texture can be done using a texturometer. The basic system of texturometer is that of a plunger, which compresses the sample at a standard rate of '43' bites per minute, the forces developed being measured by means of a strain gauge. It is claimed that with the texturometer all the parameters such as hardness cohesiveness, viscosity and elasticity can be measured. Quantitative data can be obtained from taste panels for a correlation with objective parameters. It is very important to keep a clear distinction between what is perceived subjectively and what is measured objectively.

Sensory tests are the best way by which one can get a meaningful evaluation of texture as a whole. It is always the consumers judgment, which will finally decide whether the texture of a food is good, bad or indifferent. Processing techniques like blanching, canning and cooking involve high temperature treatment, which often cause extensive changes in texture. The texture of the frozen product on thawing depends upon the freezing rate, time, temperature and relative humidity of the storage, the conditions of thawing and the composition of food. Texture deterioration, which occurs in fish upon freezing, is due to denaturation of protein.

Interaction of Taste, Odour, Texture and Flavour

Once taken into mouth, foods can make at least dual and usually a multiple impression. The tissues of mouth, throat and nasal cavity are so innervated and so interrelated that any or all of four sensory systems operate simultaneously. The cutaneous sensations of mouth react to texture, tingle and astringency, as well as to burning or biting or cooling. Thermal effects are readily perceived. The sense of taste is also involved. However, it is olfaction, which furnishes the most elaborate experiences associated with flavour. Pain features very predominantly and is the unifying sensation among all the sensory modalities. When the intensity of the stimulus, be it visual, tactile, olfactory, thermal or gustatory, becomes too strong the receptor registers a pain. Changes in texture of a food can alter the odour and flavour characteristics by changing the rate with which the stimuli reach the gustatory and olfactory receptors.

Sensory Testing Methods

Location, laboratory layout, odour control, lighting and general comfort are the important physical conditions in setting up a laboratory. The laboratory should be located in such a place so that the majority of test subjects can reach there conveniently with minimum disturbance in normal work routines. The area should be calm to avoid disturbance of tests. The testing area should be divided into tow parts, one a work area and the other for the actual testing. Individual panel booths are essential to avoid mutual distraction among test subjects. The testing area must be kept as free as possible from odours. All materials and equipment in the test room should be either odour free or have a low odour level. The testing room should have comfortable and adequate level of illumination. There should be an atmosphere of comfort and relaxation to encourage panel members to concentrate on the testing tasks.

3.3 Selection of Test Subjects

i) Discrimination Tests

Panel members are usually required to deal with complex stimuli: hence any series of tests on simple stimuli will partially determine a person's value. It is necessary to take into consideration all the factors that may affect performance. This can be done by using representative tests on representative samples. The selection process will be started with a large group of persons, since the larger the number of candidates, the greater the probability of finding persons of superior ability. Sometimes it is possible to find persons of superior ability in unexpected quarters. Requalification of all panel members is required periodically.

The first procedure in the selection of panel members is triangle test for screening. Each person should have two trials at the same test session on each triangle test. The selection should be based on no fewer than 20 judgments per subject made on 10 different tests in 10 different sessions. The top ranking people are selected and that no one scoring less than 60 percent correct will be used.

In the second test method a series of samples that represents a range of total quality are presented for judgment to the candidates. Each candidate rates the series several times. The data for each candidate are analyzed separately.

The number of panel members most often used is ten. A panel should never include a person with less than satisfactory qualifications just to achieve a predetermined panel size. The minimum number of panel members for a given test is five. If it is possible a pool of panel members should be maintained.

ii) Preference Tests

In preference tests the criterion should be representativeness of some consumer population. In this type tests all persons who have expert knowledge of the product may be eliminated. The possibility of bias and resultant error in predicting consumer preference is much greater.

The basic factors to determine the number of subjects in preference tests are the magnitude of errors, the precision desired in the results and breadth of sampling. Panels require at least 30 numbers in usual practice. About 50 to 100 people are usually considered adequate for most of the problems handled in the laboratory. When interpreting preference tests results pay particular attention to the possible effects of various factors. These tests may not be generalized too broadly.

iii. Training of Panel Members

In the discrimination tests panel members must become thoroughly familiar with the tests with which they will be concerned. This includes complete understanding of the nature of the judgments required, of the test procedure and of the test controls, which the individual is required to maintain. Training may be continued through individual and group sessions. Training should concentrate on the members perceptual and judgmental works.

In preference tests, any attempt to alter the panel members' attitude must be carefully avoided.

Following are some general suggestions for panel members

- ☆ Do not test for 1 hour after meals
- ☆ Wait 20 minutes after smoking, chewing gum or eating or d rinking between meals.
- ☆ Do not use panel members who are ill.
- ☆ Encourage panel members to avoid eating highly spiced foods on days of tests
- ☆ Panel members should not use cosmetics when running odour tests
- ☆ In taste testing, rinse out the mouth with water just before the test.

iv. Presentation of samples

The samples shall be presented in such a manner that subject will respond only on the basis of those factors which are intrinsic to the material tested. The key is uniformity, particularly with in a given test. Important factors to consider are: quantity of sample, the containers, eating utensils and temperature.

3.4 Sensory Testing Methods

Several sensory evaluation methods have been developed. Each method has its own advantages and disadvantages. The experimenter should be familiar with the merits and demerits of each method. Most practical and efficient method should be selected depending on the situation. No single method can be used universally. There will be variation and changes in the method depending on the purpose of the test and the information needed.

There are three fundamental types of sensory testing methods:

1. Preference/acceptance tests

2. Discriminatory tests
3. Descriptive tests

Preference tests are affective tests based on a measure of preference from which relative preference can be determined. Discriminative tests are used to determine whether a difference exists between samples. The panelist does not allow his personnel likes and dislikes influencing his response. Laboratory difference panels are used to determine if there is a difference among samples. Descriptive tests are used to determine the nature and intensity of the differences.

3.4.1 Preference Tests

Preference tests include the paired comparison, the hedonic scale and ranking.

i. Paired Comparison Test

Paired comparison test used in preference testing is similar to that used for difference testing when testing preferences, the panelist is presented with two odd samples and is asked which he prefers.

Questionnaire for Paired Comparison Test

Name ______________________________ Date __________

Product ______________________________

Taste the two coded samples in the following order

317, 258

Which of these two samples do you prefer?

Comments:

ii. Hedonic Scale

The most commonly used scale for preference testing is the nine-point hedonic scale. The term "hedonic" is defined as "having to do with pleasure". Here the panelist expresses his degree of liking or disliking.

Questionnaire for Hedonic Scale

Name ______________________________ Date __________

Product ______________________________

Taste the samples and check how much you like or dislike each one.

	841	419	795
(i) Like extremely			
(ii) Like very much			
(iii) Like moderately			
(iv) Like slightly			
(v) Neither like nor dislike			

(vi) Dislike slightly
(vii) Dislike moderately
(viii) Dislike very much
(ix) Dislike extremely

The ratings are given numerical values ranging from like extremely 9, to dislike extremely 1. The results are analyzed by analysis of variance. If only two treatments are evaluated, the mean scores received by each can be compared using t-test.

iii. Ranking

When ranking for preference, the panelist is presented with coded samples to rank in order of preference.

Questionnaire for Ranking

Name ______________________________ Date __________

Product ______________________________

Please rank the following 4 samples for preference. Rank the sample you like best as first and the sample you like least as fourth.

Test the sample in the following order

	817	462	149	535
First				
Second				
Third				
Fourth				

Comments:

The results are analyzed statistically.

3.4.2 Discriminatory/Difference Tests

The important test methods used to measure subjectively the difference between samples are Triangle test, Simple paired comparisons test, Scheffe paired comparisons test, Duo-trio test, Multiple comparisons test, Ranking, Scoring and Ratio scaling.

i. Triangle Test

The panelist received three coded samples, out of which two samples are same and one is different. The panelists are asked to identify the odd sample. This is a very useful method in quality evaluation to ensure that samples from different production lots are the same. It can be used to determine if ingredient substitution or some other change in manufacturing results in a detectable difference in the product. This test is often used for selecting panel members. Analysis of the results of triangle tests is based on the probability that if there is no detectable difference, the odd sample will be selected by chance one-third of the time. Standard statistical procedures are now available to determine the levels of significance. As the number of judgments increases the percentage of correct responses required for significant differences decreases. The results of triangle tests indicate whether there is a detectable difference or not between two samples. Higher levels of significance do not indicate that the difference is greater.

There are different types of questionnaire for triangle test. One of the model questionnaires is given below:

Triangle Test

Name ______________________________ Date __________

Product ____

The three samples are given 3 different 3 digit code numbers Two of the three samples are identical and the third is different.

1. **Test the samples in the order indicated and identify the odd samples.**

Code	Check odd sample
368	–
821	–
657	–

2. **Indicate the degree of difference between the duplicate samples and odd sample.**

   Slight

   Moderate

   Much

   Extreme

3. **Acceptability**

   Odd sample more acceptable

   Duplicates more acceptable

4. **Comments**

   A typical example of a triangle test, which can be used to find a detectable difference in a fish processed under two different sets of conditions, is given below. Two samples A and B from two different processing conditions are prepared and presented for triangle test. Each sample is divided into two lots. The samples are presented in coded dishes to 12 judges. Each judge received 3 coded samples. Six judges test two samples from the first treatment and the other from the second. The other six judges receive two samples from second treatment and one from the first. Because of the balanced presentation, two code numbers are assigned to each treatment. The order of presentation of the sample is randomized for each judge. Putting code numbers in appropriate order on the score sheet indicates the order in which each judge should test the samples.

The significant level is first ascertained from the chart. To find out the degree of difference, if there is a significant difference of samples between treatments, only the judgments of correct assessments are taken. The judgments are multiplied by appropriate value *i.e.* 1 for slight, 2 for moderate 3, for much and 4 for extreme and the total value is divided by the number of judges, which gives the correct degree of difference. The next part of the triangle test is to choose the more acceptable sample.

Table 3.1: Triangle Test; Level of Significance

Number of Judges	*Number of Correct Answers Necessary to Establish Level of Significance*		
	5 per cent	*1 per cent*	*0.1 per cent*
7	5	6	7
8	6	7	8
9	6	7	8
10	7	8	9
11	7	8	9
12	8	9	10
13	8	9	10
14	9	10	11
15	9	10	12
16	10	11	12
17	10	11	13
18	10	12	13
19	11	12	14
20	11	13	14

ii. Simple Paired Comparisons Test

A pair of coded samples is presented for comparison on the basis of some specified characteristics. In this method the probability of a judge selecting a sample by chance is 50 per cent. Tables for the rapid analysis of paired comparison tests are now available. One Table is used for the directional difference tests when only one answer is correct. The other Table is used in preference tests, when either response can be correct. Paired comparison tests give no indication of the size of the differences between the two samples.

A sample questionnaire for simple paired comparison test is as follows:

Questionnaire for Simple Paired Comparisons Tests

Name ______________________________ Date __________

Product ______________________________

Evaluate the juiciness of the two cooked fish samples. Test the sample on the left first. Indicate which sample is more juicy.

567 893

—— ——

Comments

Fish from each treatment is served in coded dishes to the panelists. Half the panelists are asked to taste one sample first, the others to taste the second. From the chart given below the level of significance can be determined.

Table 3.2: Simple Paired Comparisons Test; Level of Significance

Number of Judgments	*Two Tail Tests: Probability Level*			*One Tail Tests: Probability Level*		
	5 per cent	*1 per cent*	*0.1 per cent*	*5 per cent*	*1 per cent*	*0.1 per cent*
5	–	–	–	5	–	–
6	–	–	–	6	–	–
7	7	–	–	7	7	–
8	8	8	–	7	8	–
9	8	9	–	8	9	–
10	9	10	–	9	10	10
11	10	11	11	9	10	11
12	10	11	12	10	11	12
13	11	12	13	10	12	13
14	12	13	14	11	12	13
15	12	13	14	12	13	14
16	13	14	15	12	14	15
17	13	15	16	13	14	16
18	14	15	17	13	15	16
19	15	16	17	15	16	18
20	15	17	18	15	17	18

a. Multiple Paired Comparisons Test

When there are more than 2 samples to be evaluated, each must be compared with every other sample. The number of pairs is determined by the formula ½ n (n-1), where n= the number of samples or treatments. The test is called multiple paired comparisons test.

The results of multiple paired comparisons cannot be analyzed by using the same statistical table as simple paired comparisons. Computer program models have been developed which give maximum likelihood estimates.

b. Scheffe Paired Comparisons Test

Scheffe modified the paired comparisons test to task the panelists to indicate the size of the difference detected. The results are analyzed by analysis of variance. An average value for each treatment is calculated.

An example of the questionnaire and analysis of results are given below:

Questionnaire for Scheffe Paired Comparisons Test

Name ______________________________ Date __________

Examine the two samples, 864 and 165, of cooked fish for juiciness.

864 is extremely more juicy than 165 __________

864 is much more juicy than 165 __________

864 is slightly more juicy than 165 __________

No Difference

165 is slightly more juicy than 864 __________

165 is much more juicy than 864 __________

165 is extremely more juicy than 864 __________

Rate of Juiciness of each Sample

864	very dry	165	very dry
	Moderately dry		moderately dry
	Slightly juicy		slightly juicy
	Moderately juicy		moderately juicy
	Very juicy		very juicy

Comments

E.g. The Scheffe paired comparisons test is used to study the effect of four different holding temperatures on the organoleptic likeness of frozen fish.

Samples from each of the four treatments are compared with every other treatment, making a total of six comparisons. Each pair is presented to six judges for evaluation according to the above questionnaire. The experimental design requires that half the judges test one sample of a pair first and that the others taste the second sample first. The order of the six pairs was randomized for each judge.

The descriptive terms of paired comparison question are assigned numerical value ranging from +3 to –3. If samples are coded 864 and 165 respectively, three judges are asked to test 864 first and the score sheet is made up

864	is extremely more juicy than	165	+3
864	is much more juicy than	165	+2
864	is slightly more juicy than	165	+1
	No difference	0	
165	is slightly more juicy that	864	–1
165	is much more juicy than	864	–2
165	is extremely more juicy than	864	–3

The other three judges are asked to test 165 first and the score sheet is made up as follows

165	is extremely more juicy than	864	+3
165	is much more juicy than	864	+2
165	is slightly more juicy than	864	+1
	No difference	0	
864	is slightly more juicy than	165	–1
864	is much more juicy than	165	–2
864	is extremely more juicy than	165	–3

The scores for each pair are tabulated, computed and the average preference is determined. The data are subjected to analysis of variance.

Paired comparisons do not give an absolute measure of quality but only a measure in relation to the samples with which it is compared.

iii. Du-trio Test

In du trio test, three samples are presented to the panel members. One sample is a reference sample and the other two are samples to be tested. One of the test samples is identical with the reference sample and the other is different. The panelist is asked to identify the odd sample. The du-trio test has the same applications as the triangle test but less efficient because the probability of selecting the correct sample by chance is 50 per cent. It can be used in place of the paired comparisons test if no characteristic is specified. In the simple paired comparisons test the panelist is asked which sample has more of some specified characteristic, whereas in the du-trio and the triangle tests panelists bases his judgment on any difference he can detect.

Questionnaire for Du-trio Test

Name ______________________________ Date __________

Product ______________________________

On your tray you have marked control sample (R) and two coded samples. One sample is identical with R and the other is different. Which of the coded sample is different from R.

Sample	Check odd sample
234	
107	

Comments

The results are analyzed statistically.

v. Multiple Comparisons Test

In multiple comparisons test, a reference sample coded R is presented along with several coded samples. The panelists compare each coded sample with the reference sample on the basis of some specified characteristics. The multiple comparison method can be used to examine effects of changing or replacing an ingredient, of packaging material, of changing process or of storage. The test can be used very efficiently to evaluate four or five samples at a time. Small differences between the sample and control can be detected. It gives information about the direction and magnitude of the difference.

Questionnaire for Multiple Comparisons Test

Name ______________________________ Date __________

You have been supplied with samples of fish curry to compare taste. A reference sample marked R has been given. Compare the coded samples with reference sample R. Taste each sample; determine whether it is tastier than, comparable to, or less tasty than the reference. Also mark the amount of difference that exists.

Sample number	624	195	794	478
Tastier than R	—	—	—	—
Equal to R	—	—	—	—
Less tastier than R	—	—	—	—
Amount of difference				
None	—	—	—	—
Slight	—	—	—	—
Moderate	—	—	—	—
Much	—	—	—	—
Extreme	—	—	—	—

Comments

Numerical scores are assigned to the ratings by the judge, with "no difference" equaling five, extremely tastier than R equaling 9, extremely poor in taste than R equaling 1. Analysis of variance of the scores is conducted and determined the difference between the samples.

vi. Scoring

In this method samples are evaluated for the intensity of some specific characteristics. Here the panelist records his judgment on a graduated scale, which is labeled with numbers or with descriptive terms. From the scores the size and direction of the differences between the samples are determined. The descriptive terms on the scale should be carefully selected and the panelist may be trained accordingly. To use the scoring effectively, all the members must be evaluating the same characteristic. The inclusion of standards at various points in the scale will help minimize panel variability.

Questionnaire for Scoring

Evaluate the given samples for rancidity. Indicate the extent of rancidity of each sample on the scales below:

914	*721*	*863*
Not rancid	Not rancid	Not rancid
Trace of rancidity	Trace of rancidity	Trace of rancidity
Slightly rancid	Slightly rancid	Slightly rancid
Rancid	Rancid	Rancid
Very rancid	Very rancid	Very rancid
Extremely rancid	Extremely rancid	Extremely rancid

Comments

Example: The scoring method can be used to determine if there is a difference in rancidity of frozen stored sardines at three different temperatures.

The ratings assigned by the panel members can be given numerical values, ranging from 0 for 'not rancid' to 5 for 'extremely rancid'. The results may be subjected to statistical analysis such as analysis of variance.

vii. Ratio-Scaling

In this method the panelists are given a series of samples that vary in one characteristic such as hardness. The panelists are instructed to assign a number, such as 50, to the first sample and each sample is rated in relation to the first. If the second sample is twice as hard as the first the panelist assign to it a value of 100, if it is half as hard, the value is 25. The number assigned to the first sample may be chosen before hand by the experimenter or the panelist. No high or low limits are specified. One aim of the method of magnitude estimation is to have the panelist measure the same way as nature measures, with ratio scale values that have no arbitrarily limited end points.

3.4.3 Descriptive Sensory Analysis

In this analysis a group of highly trained panelists examine a particular property of a product to provide a detailed descriptive evaluation of it. The most commonly known descriptive methods are the flavour profile and the texture profile. The flavour profile is the description of the taste and odour of a product. The description names are perceptible factors, the intensity of each factor, the order in which the factors are perceived, after taste, and overall impression. The texture profile is the description of the textural characteristics perceived in a product, the intensity of each factor and the order in which they are perceived. Mechanical characteristics and geometrical characteristics are described qualitatively and semi quantitatively. Hardness, fracturability, chewiness, gumminess, adhesiveness and viscosity are mechanical characteristics. Geometrical properties are grittiness, coarseness, and fibrousness.

Descriptive analysis is a valuable tool in difference testing and product development work. In this, a complete description of sample differences is provided. This guides the product developer in modifying the product characteristics to meet consumer demands.

The training of profile for panels requires considerable time. The members must possess a high degree of motivation and interest. The trained panel can provide thorough and reliable descriptions of products in a short time. Since the descriptive panel members work together as a group, forceful members could have un due influence on the other panelists and hence change the results.

i. Descriptive Analysis with Scaling

The sensory properties of the product are identified by a trained panel. Samples are made up to illustrate the different properties so that the panel agrees on the meaning of each term used. The panel members work together as group and decide the sensory properties that should be evaluated and develop the language to be used. During evaluation session the panelist work individually.

The scale used is an interval scale consisting of a horizontal line 6 inch (15 cm) long with anchor points at 0.5 inch (1.3 cm) from each end. Each anchor point is usually labeled with a word or expression. A separate line is used for each sensory property to be evaluated. Each judge records his evaluation by making a vertical line across the horizontal line at the point, which best reflects his perception of the magnitude of that property.

After the completion of the judgment, the experimenter super imposes a grid dividing the line into 60 units, to assign a number between 0 and 60 to each rating. These values are then tabulated. Each word phrase on the scale has the same meaning to each judge and that the terms on the scale represents equal sensory intervals. For *e.g.* the distance between 'slightly rancid' and 'rancid' is the same as the distance between 'very rancid' and 'extremely rancid' in the example given under "Scoring".

Descriptive analyses with unstructured scales are also in use in certain cases. The panel wishes together as a group and identifies the important characteristics of the product to be evaluated. The terms to be used at the anchor points of the lines are agreed upon. The data are tabulated and analyzed by analysis of variance.

a. Selection of Terms for Descriptive Analysis

For a descriptive test to be reliable in the long run the descriptors must be meaningful. Quite often, the panel members are made acquainted with a list of attributes previously chosen. The panel will be taught the meaning of attributes and how to use them with the aid of given references. Two newly developed techniques – free choice profiling (FCP) and repertory grid method – appear promising in terms of saving time and generating relevant attributes.

b. Free Choice Profiling

In this technique, each panel member selects attributes, develops a score card and rate samples accordingly. Members are presented with samples belonging to the category being examined and are instructed to use a common scale. A major drawback with Free Choice Profiling (FCP) approach is that untrained people may feel distressed to retrieve.

c. Repertory Grid Method

In this method, the project leader may present each panel member with two or three samples of the same class of product and ask him to describe the product in terms of differences and similarities between the samples. The panel leader must help the assessors to dissect their own perceptual space using the interview technique. The repertory of attributes collected will be analysed to focus on analogies among descriptors obtained. Both these methods are valuable instruments for term generation and time saving.

ii. Demerit System

Demerit freshness quality grading system has several unique characteristics. Here the various sensory factors associated with different organs of fish like skin, eyes, gills, belly, vent during postmortem changes are described and graded from 0 for extremely fresh and the demerit score is increases as spoilage advances. The number of scores for each factor is given based on its contribution towards spoilage. The following is a demerit system developed by Branch and Veil (1985) having a total demerit score of 39.

The advantage of such system is that the sensory examiner is exposed to minimum confusion since every description of each demerit point is very brief, usually involving one or two words. In this system it is important to ensure that all the specified questions concerning sensory characteristics will be answered. A fish sample cannot be rejected on the basis of a single criterion since no undue importance is placed on a single feature. The minor differences in the judgments in any of the criterion being assessed do not unduly influence the total score.

Table 3.3: Freshness Quality Assessment System for Fresh Fish

Factor Being Assessed	*Observed Characteristics*	*Demerit Points*
Appearance of surface	Very bright	0
	Bright	1
	Slightly dull	2
	Dull	3
Skin	Firm	0
	Soft	1
Scales	Firm	0
	Slightly loose	1
	Loose	2
Slime	Absent	0
	Slightly slimy	1
	Slimy	2
	Very slimy	3
Stiffness	Prerigor	0
	Rigor	1
	Postrigor	2
Eyes clarity	Clear	0
	Slightly cloudy	1
	Cloudy	2
Shape	Normal	0
	Slightly sunken	1
	Sunken	2
Iris	Visible	0
	Not visible	1
Blood	No blood	0
	Slightly bloody	1
	Very bloody	2
Gills Colour	Characteristic	0
	Slightly dark/slightly faded	1
	Very dark/very faded	2
Mucus	Absent	0
	Moderate	1
	Excessive	2
Smell	Fresh oily/fresh seaweedy	0
	Fishy	1
	Stale	2
	Spoiled	3

Factor Being Assessed	*Observed Characteristics*	*Demerit Points*
Belly Discolouration	Absent	0
	Detectable	1
	Moderate	2
	Excessive	3
Firmness	Firm	0
	Soft	1
	Burst	2
Vent Condition	Normal	0
	Slight break/exudes	1
	Excessive/Opening	2
Smell	Fresh	0
	Neutral	1
	Fishy	2
	Spoiled	3
Belly cavity stains	Opalescent	0
	Grayish	1
	Yellow-brown	2
Blood	Red	0
	Dark red	1
Total demerit points (0-39)	Brown	2

3.5 Sensory Evaluation of Fresh Fish

One of the most important tests of quality of fresh fish is done by the consumers when he or she buys fish over the market slab, in the fishmongers, at the super market or from the fishermen. At this stage the consumer does not have access to the sophisticated biochemical, microbiological or other techniques that can be performed in laboratories. They do not normally have at this stage the opportunity to taste the fish either. The buyer must rely entirely on his or her senses to evaluate the freshness (or degree of spoilage) of the fish. The most common senses that are used are sight, smell and touch. If the customer does not like what he or she sees, smells or feels then the purchase and sale may not take place and the fish may not be able to be sold. This brings about hardship to those involved in the fishing industry from the fisherman through the various traders involved to the retailer himself.

Let us look at the various changes that take place during spoilage, which can be recognized. It must be remembered that different species of fish have different characteristics and rates of spoilage so that it is very difficult to generalize about the changes that might be seen at each stage.

3.6 Changes during Spoilage

i. Condition of the Skin (Sight and Touch)

Fish skin has pigment cells embedded in it which when the fish is fresh give it the characteristic colours of that fish. The skin and the colours in it are generally shiny and brilliant when it is fresh often with a sheen or luster. As spoilage progresses the products of microbial spoilage (mucus) tend to change the transparency, refraction and reflection of light through the skin, and the pigment cells begin to break down. In consequence the skin becomes dull and lacks luster and the colours fade. With the build up of bacterial slime on the skin there can be a complete masking of the original colours and the skin become covered with a yellow slime.

In fish with scales there is also a tendency for the gradual deterioration of the skin structure to bring about a loosening of the scales so that eventually they can easily be rubbed from the surface.

ii. Appearance of the Eye (Sight)

Generally in very fresh fish the eyes are bright, the centre (pupil) is black and clear and the cornea is convex and protrudes from the line of the head.

As the fish spoils there are various changes that can be seen that indicate the degree of freshness. The lens is a particularly important indicator of spoilage. The contents of the lens consist of a particular type of protein. The protein remains clear only when the aqueous solution in which it is held is at a particular ionic strength. If the ionic strength changes then the proteins precipitate out and the lens becomes cloudy. When fish is stowed in ice the ice melt water leaches out the minerals in the eye, the ionic strength changes and the eye looks milky. The longer the time in ice since death the whiter the lens tissue becomes.

As the liquid is lost from inside of the eye, it looses the roundness and it gradually sinks into the cavity in the skull.

iii. Condition of the Gills (Sight and Smell)

In very fresh fish the gills are bright blood red and have a fresh sea weedy smell. As the fish spoils the colour of the blood in the gills changes depending on the storage conditions. Having started as bright red they tend to go through dull red, may become darker or lighter depending on the degree of leaching by ice melt water. Eventually they become brown, green or yellow coloured and covered with thick mucus. The mucus is an indication of bacterial build up and will be associated with strong putrid odour. Prior to the development of the strong odour associated with well spoiled fish the gills will have lost their fresh smell and gone through a series of odour often described as neutral followed by mousy, stale, fruity *etc.*

iv. Condition of the Flesh (Touch and Sight)

The flesh of very fresh fish is firm, springy and elastic. If the surface of the fish is depressed with the thump it will return immediately to its original condition when the pressure is released. While the fish is in rigor mortis the fish is so firm that it is difficult to make an impression in the flesh at all. As rigor is resolved and spoilage progresses the flesh losses its elasticity and firmness. The thumb test results in a depression which takes more time to return to normal and when the fish is very spoiled it looses its elasticity and a thumb mark remains as a depression in the skin.

3.7 Assessing Quality

By using our senses and from what experience tells us about fish spoilage it is possible for us to judge quality in a number of different ways. The first is to make a dispassionate assessment whereby one can say that the fish "is fresh", "is stale", "is of the right quality for this or that process" *etc.* This is often called an "objective" assessment because we try to remove the feeling of like or dislike from the judgment. The second is more subjective because we ask a judge to say whether the fish is to his or her liking. Such words as excellent, like, dislike, acceptable, inedible are used in a scoring system of this sort. This is often called a hedonic system and is of little use in practical quality assessment or control for scientific or commercial use, although it might be used in marketing trials or consumer acceptance tests.

The first of these two assessment methods is most often used in the scientific and commercial assessment of fish quality. In order for the method to be used in a consistent, scientific and meaningful way attempts are made to put a number or score against the changes that occur during storage and that have been outlined earlier. Because different fish spoil at different rates and different markets have different requirements it is very difficult to produce a standard scoring system that covers all fish.

For scientific measurement the characteristics are given a numerical score so that we can say that the fish has a score of X. In commercial practice and for inspection systems the fish are usually graded according to a predetermined set of criteria. The following descriptions give some examples of scoring and grading systems for different fish. It is important that criteria such as these be drawn up individually to meet particular sets of circumstances.

3.7.1 Raw Fish

General Appearance	*Score*
Eyes perfectly fresh, convex, black pupil, translucent cornea; bright red gills (colour depending on species) – no bacterial slime, outer slime water white or transparent; bright opalescent sheen, no bleaching	5
Eyes slightly sunken, grey pupil, slight opalescence of cornea; some discoloration of gills and some mucus; outer slime opaque and somewhat milky; loss of bright opalescence and some bleaching.	3

General Appearance	*Score*
Eyes sunken; milky white pupil, opaque cornea; thick knotted outer slime with some bacterial discolouration	2
Eyes completely sunken; shrunken head covered with thick yellow bacterial slime; gills showing bleaching or dark brown discolouration and covered with thick bacterial mucus; outer slime thick yellow brown; bloom completely gone; marked bleaching and shrinkage;	0
Flesh including belly flaps	
Bluish translucent flesh, no reddening along the backbone and no discoloration of the belly flaps	5
Waxy appearance, no reddening along backbone, some discoloration of belly flaps	3
Some opacity, some reddening along backbone and some discoloration of the flaps	2
Opaque flesh, very marked red or brown discoloration along backbone, and marked discoloration of the flaps.	0
Odours	
Fresh sea weedy odour	10
Loss of fresh sea weedy odour, shellfish odour	9
No odour, neutral odour	8
Slight musty, mousy, milky or caprylic acid like odour, garlic, peppery.	7
Bready, malty, beery, yeasty odours	6
Lactic acid, sour milk, or oil odours	5
Some lower fatty acid odours (*e.g.* acetic or butyric acids) grassy, 'old boot'	4
slightly sweet, fruity or choroform-like odours	
Stale cabbage water, wet matches, phosphene-like odours	2
H_2S and other sulphide odours, strong ammoniacal	1
Indole, ammonia, faecal, nauseating, putrid odours	0
Texture	
Firm, elastic to the finger touch	5
Softening of the flesh, some grittiness near tail	3
Softer flesh, definite grittiness and scales easily rubbed off the skin	2
Very soft and flabby, retains the finger indentations grittiness quite marked and flesh easily torn from the backbone	1

3.7.2 Cooked Fish

General Appearance	*Score*
Odour	
Strong seaweedy odours	10
Some loss of seaweediness	9
Lack of odour or neutral odours	8
Slight strengthening of the odour but no sour or stale odour – wood shavings, wood sap, vanillin or terpene-like odours	7
Condensed milk, caramel or toffee-like odours	6
Milk jug odours, or boiled potato or boiled clothes-like odours	5
Lactic acid and sour milk odours	4
Lower fatty acids (*e.g.* acetic or butyric acids) some greasiness or soapiness, turnipy or tallowy odours	3
Ammoniacal (trimethylamine) and some sulphide odours	2
Strong ammoniacal(trimethylamine) and some sulphide odours	1
Strong ammonical and faecal, indole and putrid odours	
Flavour	
Fresh, sweet flavour, characteristic of the species	10
Some loss of sweetness	9
Slight sweetness and loss of the flavour characteristic of the species	8
Neutral flavour, definite loss of flavour but no 'off' flavour	7
Absolutely no flavour, as if chewing cotton wool	6
Trace of 'off' flavour, some sourness but no bitterness	5
Some 'off' flavour, and some bitterness	4
Strong bitter flavour, rubber-like flavour, slight sulphide-like flavor	3
Strong bitterness, but not nauseating	1
Strong 'off' flavour of sulphides, putrid, tasted with difficulty	0

Texture Profile

Initial Characteristics

Response to the properties of the material on the first bite.

Wateriness

It is connected with the release of water on compression. This is the initial response and is to be distinguished from succulence.

- ☆ No water released but not necessarily dry
- ☆ Wet, sample releases water very readily

Firmness

The force required for compressing the material between the teeth or between tongue and palate.

- ☆ Very soft, very easily compressed
- ☆ Firm, high resistance to compression

Springiness

The ability of the material to return to its original shape after deformation. Compress the substance slightly between the molar teeth or between tongue and palate and note how the material returns to its original shape.

- ☆ Completely plastic, retains its deformed state
- ☆ Springy, returns to its original shape

3.8 Secondary Characteristics

Response to the properties of the material after chewing a few times.

Fibrousness

This is the property of separating into filamentous structural elements.

- ☆ Not fibrous
- ☆ Short fibres, almost mealy
- ☆ Long fibres

Smoothness

This is the alternative to fibrousness in describing the geometry of the structural elements

- ☆ Fibrous
- ☆ Crumbly, mealy, lumpy
- ☆ Homogeneous –smooth, pasty

Toughness

This is the resistance to breakdown on chewing to a state suitable for swallowing.

- ☆ Very tender, very easily broken down
- ☆ Very tough, needs chewing for a long time

Succulence

This is the sensation of juiciness in the mouth.

- ☆ Very dry, tends to reduce the moisture in the mouth
- ☆ Succulent, juicy, tends to increase the moisture in the mouth

Others

There is many other texture characteristics not covered by the defined scales *e.g.* tacky, oily, glutinous, brittle and crisp.

Different evaluation systems are used depending on the requirement. The most commonly used test forms are paired comparisons, rating scales, rank order, forced choice methods, threshold methods, and methods of quality attribute analysis which are described earlier.

The various aspects of sensory evaluation described above are important because they provide an insight into sensory testing and their practical value in quality evaluation and product development. Fish from unconventional sources have become increasingly important due to the search for more aquatic products because of the health benefits attached to it. Sensory evaluation provides a description of the various sensory qualities these unconventional products have and how it can be modified so that it is acceptable to the consumers. A thorough knowledge of sensory qualities and various testing methods are very important in new product development and hence reducing post harvest fishery losses.

It shall not forget that the success of sensory evaluation depends on the individuals involved and their ability to make meaningful contributions to the decision making process.

3.9 Physical (Instrumental) Method for Assessing Seafood Quality

a. Freshness Meters

Based on the changes taking place in the electrical properties of fishmuscle (such as conductance and capacitance) a freshness meter has beendeveloped at Torry Research Station (U.K.) known as Torrymeter (TM) which has readings from 0 to 16. A similar meter with a wider range (0 to 100) has been developed in Germany, known as Intelectron Fish Tester (IFT). These meters give quick and reliable indication of fresh fish quality in tropical fish. In GR Torrymeter, highest value (16) is obtained for very fresh fish and the readings decrease with spoilage and is characteristic of the species. The RT-Freshness Grader developed by Iceland is a commercialsuccess in this series. The grade measurement is faster and correlates well with sensory (odour) and chemical (TMA-value) assessment of freshness.

b. Texture Measurement

The texture of the fish is often a good measure of the quality and might be determined most often in shear or compression test. The Universal Testing Machine (UTM) and other commercial texturometers (RHEO TEX,Japan) are used to measure objective textural quality.

c. Electronic Nose

An electronic nose is a device intended to detect odors or flavors. Over the last decade, "electronic sensing" or "e-sensing" technologies have undergone important developments from a technical and commercial point of view. The expression "electronic sensing" refers to the capability of reproducing human senses using sensor arrays and pattern recognition systems. Since 1982,[2] research has been conducted to develop technologies, commonly referred to as electronic noses, that could detect and recognize odors and flavors. The stages of the recognition process are similar to human olfaction and are performed for identification, comparison, quantification and other applications, including data storage and retrieval. However, hedonic evaluation is a specificity of the human nose given that it is related to subjective opinions. These devices have undergone much development and are now used to fulfill industrial needs.

The electronic nose was developed in order to mimic human olfaction that functions as a non-separative mechanism: *i.e.* an odor/flavor is perceived as a global fingerprint. Essentially the instrument consists of head space sampling, sensor array, and pattern recognition modules, to generate signal pattern that are used for characterizing odors. Electronic noses include three major parts: a sample delivery system, a detection system, a computing system.

3.10 Chemical/Biochemical Quality Evaluation of Seafood

The eating quality of fish is one of the important attributes that influence the acceptability of fish as food to the consumer. If the fish is not good to eat the consumer will not buy it again. The quality of fish begins to deteriorate soon after the catch, and hence it is essential to process the catch as quickly as possible to maintain its freshness and to prevent deterioration. Quality of the catch can be assured and maintained if the same can be measured objectively at all stages from catch to consumption.

Presently the most commonly used method for quality assessment of raw fish is sensory evaluation. Simplicity and rapidness are the advantages of the system but this lack objectivity. Therefore, many attempts were made to find out alternate methodologies for fish quality evaluation such as the determination of volatile compounds by using sensors and electronic nose.

During spoilage of fish a number of chemical reactions take place in the fish muscle. In the chemical assessment of quality, various compounds formed during these reaction are quantitatively determined and correlated with sensory characteristics. The compounds are produced in fish muscle by autolytic enzymes, putrefactive microorganisms or by chemical reactions such as lipid oxidation. During spoilage these compounds gradually get accumulated in the flesh and hence their determination provides a measure of the progress of spoilage. Following are the groups of compounds formed during spoilage.

i. Volatile Bases

Basic nitrogenous compounds such as ammonia, trimethylamine oxide (TMAO), Trimethylamine (TMA) and Dimethlamine

ii. Nucleotides

Degradation products from Adenosine triphosphate (ATP) *e.g.* Inosine monophosphate and hypoxanthine

iii. Lipid Oxidation Products

Peroxides, hydroperoxides and aldehydes.

3.10.1 Volatile BNases

i. Total Volatile Bases (TVB)

Total volatile bases refer to all the volatile basic compounds mainly trimethylamine and ammonia. The Total Volatile Base Nitrogen (TVBN) value along with TMA is the most common index of quality universally used for deciding the state of freshness of fish. Some TVB will be naturally present in very fresh fish (<20mg per cent). Volatile bases are produced by spoilage bacteria in fish. 35-40 mg TVBN/100g of muscle is usually regarded as the limit of acceptability beyond which the fish can be regarded as spoiled.

ii. Trimethylamine (TMA)

Trimethylamine (TMA) is used to assess the freshness in marine fish. This index is not suitable for freshwater fish and heat treated fish products. TMA is derived from trimethylamineoxide (TMAO) which is critical for osmo regulation in marine fish. TMAO is reduced by bacterial enzymes to TMA while fish endogenous enzymes reduce TMAO to DMA and then to formaldehyde. The DMA only increases when bacterial growth is halted by freezing. For this reason, TMA is related to bacterial spoilage of fresh fish and DMA can be used as a quality marker for frozen fish. However, no limits for DMA content has been proposed by European Unions. There are no toxicological problems directly related to these volatile amines. Trimelthylamine level in fish muscle is used as more specific index of bacterial spoilage. A level of 10-15 mg TMA-N/100g muscle is considered as the limit of acceptability. This level increases with storage time at ambient temperature in ice or in refrigerated sea water and hence forms a good index of spoilage.

Determination of TVB-N and TMA-N

TVB-N and TMA-N can be accurately determined by Conway micro diffusion method. Conventional standardization method can also be adopted for the determination of TVBN and TMAN. Here the TCA extract of the muscle is made alkaline with NaOH in a steam distillation apparatus and the evolving bases are absorbed in 2 per cent boric acid. The volatile bases are determined by tiration with standard acid. It is suggested that level of TMA in fresh fish expressed as nitrogen (N), TMA should be lower than 10-15 mg/100g of fish, but a maximum level of 5 g of TMA-N/100g has been proposed by European Union.

3.10.2 Nucleotide Based Method

Chemical quality tests must accurately reflect the edibility or the sensory quality of the product. The nucleotide degradation products especially inosine monophosphate (IMP), hypoxantine (H_x) or K value clearly reflects the quality loss in fish. The presence of higher levels of IMP in the muscle indicates relatively high quality, whereas accumulation of inosine and hypoxathine is an indicator of poor quality.

Adenosine Triphosphate (ATP) is an important in the storing energy in most living organisms. When fish die ATP is broken down over a period of days by enzymes present in the flesh, to different substances. The final stages of this process is the formation of a compound called hypoxamthine, which gradually increase with time and can be used as a measure of quality of fish. The rate of accumulation of hypoxathine is not the same in all species. The amount of hypoxathine present is measured by the enzymic method or by High Pressure Liquid Chromatography (HPLC) method. K value is one of the most appropriate indicators of freshness. Like hypoxathine, the K value measures the extent of the breakdown of ATP. It is the percentage of the intact ATP present at death that has been converted by enzymic action into hypoxathine and its immediate precursor called inosine in the chain of decomposition of ATP. HPLC method is used to determine the K value. K value as an index of estimating the freshness of fish has become widely used in Japan.

K value can be defined as,

$$K = \frac{HxR + Hx}{HxR + Hx + ATP + ADP + AMP + IMP} \times 100$$

where,

H_xR = Inosine

H_x = Hypoxathine

ATP = Adenosines triphosphate

ADP = Adenosines diphosphate

AMP = Adenosines monophosphate

IMP = Iosines monophosphate

Since the freshness of fish deteriorates rapidly, freshness may be considered as synonym for quality. Freshness determines the quality of food fish not only those consumed raw or used in home cooking, but also those to be processed

Total CaATPase activity was proposed as an index for estimating quality of surimi. This index is based on the degree of insolubility of myosin fraction.

3.10.3 Rancidity Tests

i. Peroxide Value

It measures peroxides and hydroperoxides formed during the oxidation of fat in fish muscle. The most common method is based on iodometric titration which measures the iodine produced from potassium iodide (KI) by the peroxide present in fat. PV is a good guide to quality of fat. Fresh oil should have PV 1 mg.oxygen/kg. On storage it may increase to 10 mg/kg.

The oil from fish muscle is extracted into chloroform after dehydrating the muscle with anhydrous sodium sulphate. The solution is filtered. 10 ml of the chloroform extract is taken in a conical flask and mixed with 3 - 5 ml aldehyde free acetic acid. One ml of saturated KI solution is added and the solution kept in the dark for 10 minutes. It is then titrated against standard N/500 thiosulphate solution using starch as indicator.

PV = ml of N/500 thio/g of fat

ii. Thiobarbituric Acid Value (TBA Value)

TBA measures the malonaldehyde produced during fat oxidation. TBA reacts specifically with malodadehyde to give a red cromogen which can be determined spectrophotometrically. The test can be performed in two ways either directly in food followed by steam distillation and allowing the distillate to react with TBA reagent or by preparing an extract of the muscle followed by colour development. The absence of the red coloured pigment is measured at 538 nm against a reagent blank. The TBA number is calculated as milligram of malonaldehyde per kg of sample, which is equal to 7.8 times of the optical density.

TBA number - OD x 7.8

The PV is a measure of the first stage of oxidative rancidity and TBA is the second. It can assessed that if PV is above 10 - 20 mg of O_2/kg of sample then TBA will be above 1-2 MA/kg of sample of smell and taste rancid.

iii. Aspartylaminotransferase

During faulty freezing and frozen storage all damage due to ice crystal formation occurs. The enzyme aspartylaminotransferase is released into the thaw drip during thawing. Hence the content of this enzyme is a measure of the extent of cell damage.

3.10.4 Ammonia

Bacteria can generate small amounts of ammonia, in spoiling fish, mainly from free amino acids. The amount of ammonia can give an indication, though not accurate one, of the extent of spoilage. Much larger amounts of ammonia are produced during spoilage of elasmobranchs fishes like skate and dog fish because they have large amount of urea in their flesh. Shellfish also may develop more ammonia than most marine fish and at an early stage.

3.10.5 Biogenic Amines as an Index of Spoilage

Biogenic amines are nonvolatile compounds, which are found at very low level in fresh fish and their accumulation is related to bacterial spoilage. Histamine and other biogenic amines have been proposed as markers to evaluate fish freshness. Important biogenic amines are histamine, cadaverine, putreseine, tyramine, tryptamine, spermine and spremidine. Histamine is known to be the causative factor of scombroid poisoning and level greater than 5 mg/100g imply fish decomposition. In Mackerel, tuna, bonito, herring and sardine the production of toxic amines is an indication of fish spoilage. Dark fleshed fish have high histidine content and spoilage organisms convert it to histamine.

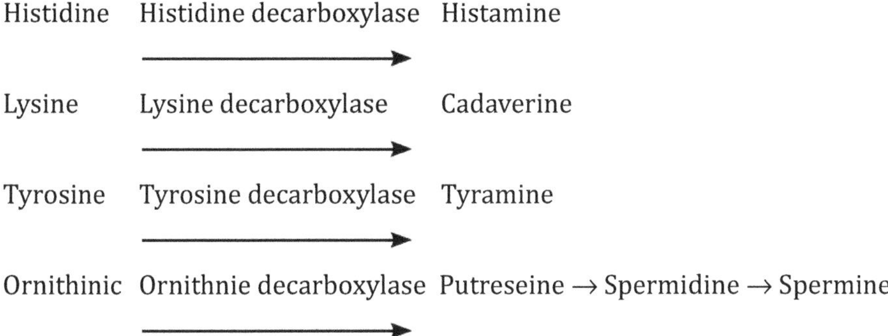

Biogenic amines resulting from fish spoilage are thermally stable and thus can be used as indicator of poor quality raw material in preserved products. Evidence also showed that Cadaverine and Putreseine are potentiators of histamine and hence potential to cause illness even in the absence of histamine. The direct precursors of histamine, cadaverine and putreseine are histidine, lysine and ornithine respectively. The break down of proteins in the latter stages of spoilage contributes additional amino acids for continued bacterial activity. Fish proteolytic enzymes and peptidases play important roles in the postmortem biochemical spoilage of fish during initial days. Incidence of histamine poisoning after eating fish are mainly due to poor quality of raw material or poor handling.

Biogenic amines in fish are also related to various health effects such as histamine intoxication or food intolerance crisis *etc.* In general, adverse effects of biogenic amines in fish are related to high content of these compounds; but low levels may also be hazardous for some people. Thus during histamine intoxication and a monamine oxidase inhibitor therapy is strongly recommended to avoid hypertension crisis. Although the biogenic amines have been associated with fish spoilage legal limit has been established for histamine only. The European Union set a maximum average content of 100 mg/kg fish for canned products and trace for ripened products. The US Food and Drug administration lowered the limit from 100 to 50 mg/kg, recommending that not only histamine level but also other biogenic amine content had to be taken into account. India has also limited maximum permissible level for histamine content in frozen fish 200 mg/kg. Less than 5 mg/kg is considered safe for consumption; 5–20 mg/kg is safe; 20–100 mg/kg is probably safe and > 100 mg/kg is toxic and unsafe for consumption.

Methods of Detection of Histamine and other Biogenic Amines

The analysis of histamine in the ingested food is critical to the confirmed diagnosis of histamine poisoning. A variety of methods are available for the analysis of histamine and biogenic amine in foods ranging from bioassay to chemical assays, which include flurometric, enzymatic and chromatographic procedures. Chromatographic and high pressure liquid chromatography methods (HPLC) are considered to be suitable for determining biogenic amine in sea food. Histamine can be determined by the method of Hardy and Smith (1976) An enzyme-based colourimetric method for histamine in fishmeal was reported by Albredet piz *et al.* (2000).

Regulatory limits of histamine in fish and fishery products has been established in several countries. Maximum permissible limits for other biogenic amine such as putreseine and tyramine were not prescribed by any regulatory agencies in US, EU and Japan.

It was reported that cadaverine and putreseine could be used as freshness indices for fish and shellfish respectively. Fish and fishery products containing cadverine below 15 mg per cent are considered as good for consumption, 15-20 mg per cent indicates potential decomposition and over 20 per cent advanced decomposition.

Table 3.4: Permitted Levels of Limits for Histamine in Fishery Products

Country	*Limit*
USFDA	50 mg per cent (hazard action level)
	10-20 mg per cent (defect action level)
EEC	10 mg per cent (defect action level)
	20 mg per cent maximum allowable unit
Canada	> 10 mg per cent indicator composition
	10 – 20 mg per cent (defect action level)
Germany	20 mg per cent
Denmark	30 mg per cent
India	20 mg per cent
Sweden	20 mg per cent

3.10.6 Indole as an Index of Quality

Indole production is an indicator of spoilage of shrimp. Indole is a degradation production of tryptophan. Indole is currently used by the USFDA to validate the sensory evaluation of shrimp decomposition. Shrimp with indole content < 25 µg/100g is organoleptically acceptable. Indole is usually determined by spectroflurometric or spectrophotometric method of AOAC.

Indole is extracted with light petroleum from trichloro acetic acid – precipitated shrimp muscle. The extracted indole soluble in light petroleum is reacted and re-extracted with Ehrlich's reagent.

3.11 Microbiological Quality Evaluation of Seafood

Since fish is harvested from natural water bodies including farms, it harbours a number of microorganisms found in the environment from where it is caught. These native microorganisms may include fish spoilage bacteria as well as certain pathogens of aquatic origin. In addition to these inherent microorganisms, the fish can get contaminated with other microorganisms during handling, transportation and processing, right from the point of catch to the end product. These microorganisms include both pathogenic and non-pathogenic bacteria. The pathogenic microorganisms can be hazardous to the health of the consumer. The most important pathogens which gain entry into the fish during handling, transportation and processing are *Salmonella, Vibrio cholerae, Staphylococcus aureus* and *Listeria monocytogenes.* In addition, enteropathogenic *Escherichia coli, Clostridium perfringens* and *Bacillus cereus* may also gain entry to the fish.

In order to determine the suitability of fish as well as shellfish, the assessment of these microbial qualities is necessary. The microbial parameters that are generally assessed to determine their consumer acceptability and safety are the following.

1. Total plate count (TPC)
2. Total *Enterobacteriaceae* count (includes all coliforms, *Salmonella and Shigella*)
3. *Escherichia coli* (*E. coli*)
4. *Staphylococcus aureus*
5. *Faecal streptococci*

The level of the total bacteria (TPC) will indicate the freshness of the products, as well as their potential shelf life. The type of bacteria in the product can give information regarding how the product has been handled or processed. A high bacterial count indicates the level of contamination of the product, conditions of storage, the extent of spoilage, *etc.* Hence TPC gives an overall picture of the quality of the product. A TPC of 10^6/g or above is considered as a proof of poor quality of the product.

The TPC can be determined by microscopic or cultural method. The cultural method is preferred for determination of TPC because it gives an estimate of viable (live) cells. A suitable culture media, like Tryptone Glucose Agar (TGA) or plate count agar is generally used for determination of TPC. Live bacterial cells are capable of forming colonies on a suitable solid medium. This property of the bacteria is made use of in the determination of TPC.

Total enterobacteriacea count and total *faecal streptococci* count gives an indication to whether there was any faecal contamination in the fish at any stage; *E. coli* is a direct faecal indicator organism. *Staphylococcus aureus* can arise mainly from the human handlers; *Salmonella* and *V. cholerae* are pathogenic organisms, implicated in food-borne infections.

Table 3.5 gives the various microbial parameters to be studied for fish and fish products and the media used for their sampling.

Table 3.5: Microbial Parameter and the Media for Sampling

Parameter	*Medium*
Total plate count (TPC)	Tryptone Glucose Agar (TGA)
Total *Enterobacteriaceae* count (includes all coliforms, *Salmonella and Shigella*)	Violet Red Bile Glucose Agar (VRBGA)
Escherichia coli (*E. coli*)	Tergitol 7 Agar (T 7)
Staphylococcus aureus	Baird Parker Medium (BP)
Faecal streptococci	Kenner Faecal Streptococci Agar (KF)
Salmonella	A set of media
Vibrio cholera	A set of media

Chapter 4

Concept of Quality Management

The traditional quality control program was based on establishing effective hygiene control. Confirmation of safety and identification of potential problems was obtained by end-product testing. Control of hygiene was ensured by inspection of facilities to ensure adherence to established and generally accepted Codes of Good Hygiene Practices (GHP) and of Good Manufacturing Practices (GMP).

4.1 Traditional Quality Control

Codes of GHP/GMP

Inspection of facilities and operations

End-product testing

Codes of GHP/GMP are still the basis of food hygiene. However, codes –although being essential – only provide for the general requirements without considering thespecific requirements of the food and the processing of specific foods. Also the requirements are often stated in very imprecise terms such as "satisfactory", "adequate", "acceptable", "suitable", "if necessary", "as soon as possible" *etc.* This lack of specifics leaves the interpretation to the inspector, who may place too much emphasis on relatively un important matters. He may fail indistinguishing between "what is nice and what is necessary" and consequently increase the cost of the programme without reducing the hazards. Perhaps one of the most common mistakes that many inspection services and some foodcompanies make is to rely on end-product testing. Very often this has been the only quality and safety assurance system applied. Samples have been taken randomly from the day's production, and examined in detail in the laboratory. There are several problems related to this procedure:

It is costly. A well equipped laboratory will be needed as well as trained personnel. The running costs of a laboratory is high. Also, the cost of products "lost" to testing may be very high; the results are retrospective, and all cost and expenses have already been incurred if any hazards are identified in the end-product testing programme. What is needed is a preventive system, where safety hazards are anticipated and safety is built into the product right from the start; it may take several days before results from end-product testing are available; the chances of finding a hazard will be variable, but most often very low (see below).

Nevertheless, the hard work of sampling and testing will give a sensation of "being in control" and create a strong but false sense of security. It is important to understand the ineffectiveness and limitations in using end-product sampling and testing to ensure product safety. In most cases there is no test that give an absolutely accurate result with no false positives and no false negatives. This is certainly the case for all microbiological testing. Furthermore, there are the principles of sampling and the concept of probability to consider.

4.2 Modern Safety and Quality Assurance Methods and Systems

To the uninitiated, and also the initiated, there may seem to be a whole host of different options or methods for ensuring the safety and quality of food products. The situation is not helped by the acronyms arising from these methods *i.e.* ISO, GMP, GHP, HACCP, TQM, *etc.* seeming to have a life of their own and coming into modern usage as words in themselves, and sometimes used without an understanding of what they mean. This brief section tries to succinctly define what each of these methods are and what they were designed to achieve.

Management concern - items to be managed

1. Technical : Intrinsic quality of fish (taste, smell and texture); safety; spoilage/freshness; grading; packaging; nutritional; authenticity; shelf life, *etc.*
2. Managerial : Administrative systems; customer relations; promotion; delivery commitments; invoicing and payment, *etc.*
3. Environmental : Waste and water management; noise pollution; odours; pollutants, *etc.*

4.3 Methods to Manage Quality and Safety

Below are listed the most well known methods to manage quality and/or safety, and these are bebriefly discussed individually.

1. Good Hygienic Practices (GHP)/Good Manufacturing Practice (GMP) or Sanitation
2. Standard Sanitation Operating Procedures (SSOP) or prerequisite programmes
3. Hazard Analysis Critical Control Point (HACCP)

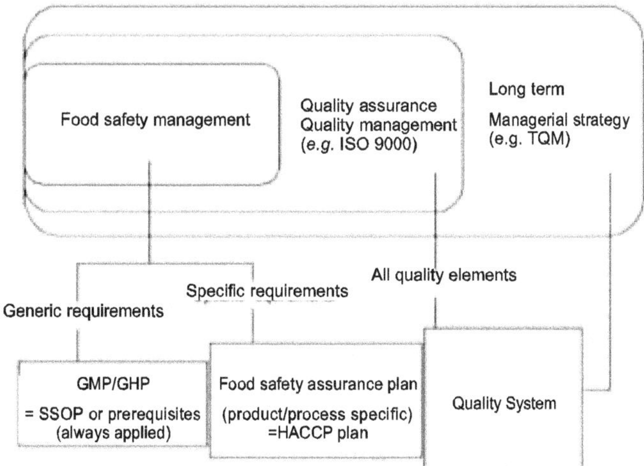

Figure 4.1: Food Safety Tools: An Integrated Approach.

4. Quality Control (QC)
5. Quality Assurance (QA)/Quality Management (QM) - ISO standards
6. Quality Systems
7. Total Quality Management (TQM).

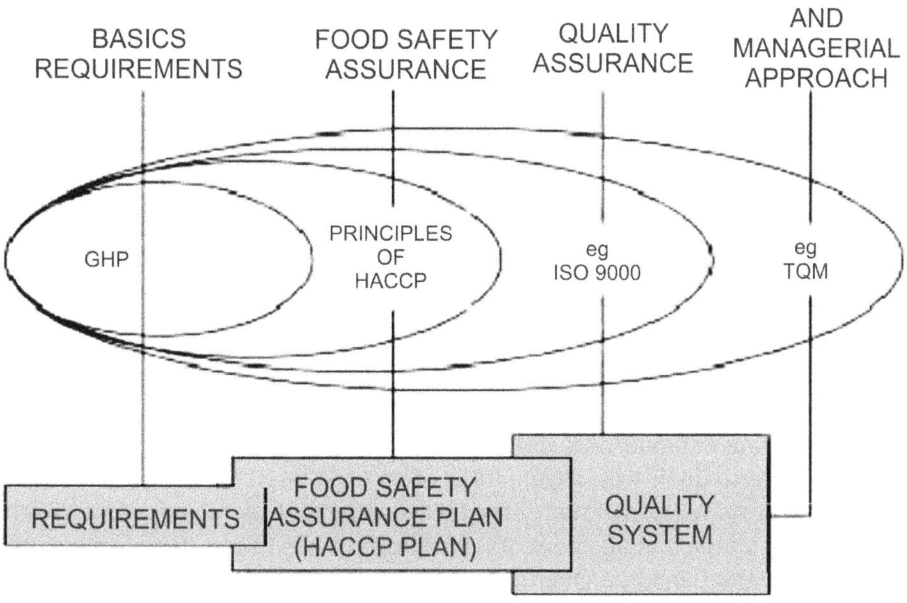

Figure 4.2: Management of Quality and Safety.

i. Good Hygienic Practices/Good Manufacturing Practices

The terms GHP and GMP basically refer to measures and requirements which any establishment should meet to produce safe food. These requirements are prerequisites to other and more specific approaches such as HACCP, and are Procedures (SSOP) has also been used in the US to encompass basically the same issues, *i.e.* best practices. often now called prerequisite programmes. In recent years the term Standard Sanitary Operating procedure

ii. Hazard Analysis Critical Control Point

Hazard Analysis Critical Control Point (HACCP) is a systematic approach which identifies, evaluates, and controls hazards which are significant for food safety (CAC, 1997). HACCP is discussed in great detail throughout this book. In the context of this section, HACCP ensures food safety through an approach that builds upon foundations provided by good manufacturing practice. It identifies the points in the food production process that require constant control and monitoring to make sure the process stays within identified limits. Statistical Process Control systems are relevant to this operation.

HACCP is legislated in many countries, including the USA and the European Union. The combination of GHP/GMP and HACCP is particularly beneficial. In that the efficient application of GHP/GMP allows HACCP to focus on the true critical determinants of safety.

iii. Quality Control (QC)

Can be defined as the operational techniques and activities that are used to fulfil quality requirements (Jouve *et al.,* 1998) It is an important subset of any quality assurance system and is an active process that monitors and, if necessary, modifies the production system so as to consistently achieve the required quality. It can be argued that QC is used as part of the HACCP system, in terms of monitoring thecritical control points in the HACCP plan. However, traditional QC is much broader than purely this focus on critical control points for safety systems. The pitfalls of relying on QC procedures,more importantly as end product testing.

iv. Quality Qssurance/Quality Management

This can be defined as all the activities and functions concerned with the attainment of quality in a company. In a total system, this would include the technical, managerial and environmental aspects as alluded to above. The best known of the quality assurance standards is ISO 9000 and for environmental management, ISO 14000. The term quality management is often used interchangeably with quality assurance. In the seafood industry, the term quality management has been used to focus mostly on the management of the technical aspects of quality in a company, for instance, the Canadian Quality Management Programme which is based on HACCP but covers other technical issues such as labelling.

ISO Standards

The International Organization for Standardization (ISO) in Geneva is a worldwide federation of national standards bodies from more than 140 countries.

The mission of ISO is to promote the development of standardization and related activities in the world with a view to facilitating the international exchange of goods and services, and to developing cooperation in the spheres of intellectual, scientific, technological and economic activity. ISO's work results in international agreements which are published as International Standards. The vast majority of ISO standards are highly specific to a particular product, material, or process. However, two standards, ISO 9000 and ISO 14000, mentioned above, are known as generic management system standards. Over half a million ISO 9000 certificates have been awarded in 161 countries and economies around the world and in 2001 alone over 100 000 certificates were awarded, 43 per cent of which were the new ISO 9001:2000 certificate. Historically, the ISO 9000 series of standards of relevance to the seafood industry included:

- ISO 9001 Quality systems - Model for quality assurance in design/ development, production, installation and servicing
- ISO 9002 Quality systems - Model for quality assurance in production and installation.

More recently, the new ISO 9001:2000 certificate is the only ISO 9000 standard against whose requirements a quality system can be certified by an external agency and replaces the old ISO 9001, 9002 and 9003 with one standard.

It is important to note that the ISO 9000 standards relate to quality management with customer satisfaction as the end point, and that they do not specifically refer to technical processes only. ISO 9000 gives an assurance to a customer that the company has developed procedures (and adheres to them) for all aspects of the company's business. ISO 14000 is primarily concerned with environmental management. Introduced much later than the ISO 9000 series, there are now over 35 000 ISO 14000 certificates awarded in 112 countries or economies of the world. During 2001, nearly 14 000 certificates were awarded, around 40 per cent of the total awarded since the introduction of the standard.

In most countries, implementation of ISO 9000 quality management systems or ISO 14000 environmental systems are voluntary.

v. Quality Systems

This term covers organizational structure, responsibilities, procedures, processes and the resources needed to implement comprehensive quality management (Jouve *et al.,* 1998). They are intended to cover all quality elements. Within the framework of a quality system, the prerequisite programme and HACCP provides the approach to food safety.

vi. Total Quality Management (TQM)

TQM is an organization's management approach, centred on quality and based on the participation of all its members and aimed at long-term success through customer satisfaction and benefits to the members of the organization and to society (Jouve *et al.,* 1998). Thus TQM represents the organizations' "cultural" approach and together with the quality systems provides the philosophy, culture and discipline necessary to commit everybody in the organization to achieve all the managerial objectives related to quality.

The Concept of Total Quality Management in Seafood Industry

Sea food occupies a unique position as food material for man. A variety of nutrients both major and minor are available in almost all fish and fishery products. The high nutritional quality and easy digestibility of fish make it a favorable food of almost all living organisms including bacteria. If handled carelessly sea food can become a source of different kinds of organisms like *V. cholera, S. typhi, L. monocytogenes, V. parahaemolyticus, S. aureus etc.* which can cause various kinds of infectious diseases and food poisoning in man.

The variety and species of aquatic organisms used as food are too many. Some of these organisms by virtue of their genetic make up or food habit are found to contain some toxic substances. Thus ciguatoxin, paralytic shellfish poison, diarrhetic shellfish poison, amnesic shellfish poison *etc.* are species related toxins which are health hazards encountered in sea food by man. Similarly scombroid fishes on temperature abuse produce histamine and cause the commonly reported scrombrotoxin poisoning. Some species are known to accumulate certain toxic residues from the eco system to toxic levels. The cephalopods squid and cuttle fish are reported to accumulate Lead and Cadmium depending on pollution of the environment as well as age of the organism. On a similar fashion large fishes like tuna and marlin are known to accumulate mercury to toxic levels with increase in age and size. Fish and shellfish raised by farming are likely to concentrate environmental contaminants and aquaculture drugs like pesticide residues and antibiotic residues. All these residues pose various kinds of health risks to consumers.

In spite of all these probable health hazards fish and shell fish continue to be in great demand as a food material. To avoid public health problems in using fish and shell fish as a food of mass consumption several quality assurance programmes have been evolved. Various kinds of quality standards like Codex standards, US FDA standards, BIS standards *etc.* and HACCP system of USA, European Commission Norms, QMP of Canada and TQM of Japan aim at ensuring safety and quality of fish and fishery products. Even though TQM of Japan aimed at total quality management, it failed to ensure both safety and quality, probably due to certain lacuna like calibration, good laboratory practice, good personnel policy *etc.* There were several attempts to improve the TQM concept of Japan to make it suitable to tackle all problems of safety and quality of a given product.

Today Total Quality Management (TQM) is a widely used technique for quality assurance in a wide range of production industries. The use of TQM is gaining more and more importance in seafood industry perhaps due to the higher incidence of health risks/hazards in fish and fishery products.

Even though widely used, depending on various factors the approach and concepts for implementation of TQM vary from industry to industry and from person to person. Obviously there is a need for consolidation of all relevant aspects to attain uniformly assured quality for products of mass consumption. What follows is an attempt to consolidate systematically all necessary steps to prevent/eliminate health risks/hazards from all possible sources in a food processing industry.

The first step in deciding a procedure for TQM is to list out all possible sources of health risks/hazards. In any system of food production the risks/hazards depend upon one or a combination of the following:

1. Raw material
2. Production Process
3. Production facility- Plant and Machinery
4. Personnel involved in Production
5. Cleanliness of direct/indirect food contact surfaces
6. Pests
7. Risk/hazard monitoring facilities - Laboratory
8. Personnel involved in production and quality checks

vii. Good Manufacturing Practices in Fish Handling

Protection of consumer against food borne diseases and maintenance of manufacturer's reputation are mainly based on the levels of sanitation in the processing establishments. The size of the individual production unit in this country varies from small processing units running on marginal profits to well established factories adopting mechanisation in some of the stages of processing. This unevenness within the industry is often reflected in varied approaches towards problems of sanitation. Flesh of live fish is free from bacteria. When the fish is dead, bacteria present on the skin,gills and in the intestine will start acting on the muscle simultaneously producing many metabolites having offensive smell. At this stage the fish is said to be spoiled and it is no more edible. Elevated temperatures, poor sanitary conditions of the factory premises and inadequate hygiene of the workers will speed up the fish spoilage. Any amount of money spent in the factory will result only in diminishing returns unless a well organised programme of sanitation is charted out and effectively implemented. The responsibility of such, a programme should lie with the management, the technologists and other quality assurance personnel in the factory. As it is not possible for the management to take personal cognizance of all the parameters to maintain a sanitary plant, this responsibility and authority must be delegated. The management may also see that the codes of sanitation are strictly adhered to in the construction and layout of factories and the facilities required are offered in liberal terms.

Care on Board Fishing Vessels

1. Fishing shall not be done from polluted waters.
2. Fish hold and boat-deck shall be constructed in such a way that they can be easily cleaned and disinfected.
3. All fish handling surfaces shall be smooth, non-corrosive and free from cracks and crevices.
4. It is not advisable to use bamboo baskets, wooden boxes and similar containers that are difficult to be cleaned.
5. Between trips, all the fish contact surfaces shall be cleaned using a suitable detergent like teepol followed by disinfection using chlorine of 100 ppm. strength giving a contact time of 15 minutes.
6. Fishing boats shall carry ice prepared from potable water/sanitary sea water.
7. Ice used during one fishing trip shall not be used for the subsequent trips.
8. After each haul, the catch may be washed with potable water chlorinated to a residual level of 10 ppm.
9. Immediately after washing, the fish shall be mixed with ice in theratio 1 : 1 and stored in clean containers.
10. The boxes used for packing fish shall be of polythene or any other suitable synthetic material that can be easily cleaned and disinfected.
11. The boxes shall be designed and arranged in such a way that the ice-melt water carrying blood, dirt and slime from one box shall not fall on the box underneath.
12. There shall be proper drainage facilities to take away the ice-meltwater.
13. If fishes are handled on board using showels, care may be taken so that the flesh is not damaged.
14. Walking or standing over fishes is not advisable.
15. Immediately after storing the fishes of each haul, the deck shall be cleaned and disinfected.

Care during Unloading

1. During unloading fishes, care shall be taken so that the fishes are not exposed to adverse elements of nature.
2. Washing the catch using coastal seawater is dangerous.
3. Sorting the catch on sea beaches shall be discouraged.
4. Fishes de-iced for weighing shall be re-iced and chilled below 2°C as quickly as possible.

5. Proper precautions have to be taken to prevent the entry of flies, crows, cats, dogs, rodents *etc.* in the premises where fishes are unloaded, weighed and re-iced.
6. All the containers/contact surfaces used for unloading and weighing shall be cleaned and disinfected immediately.

Care during Transportation

1. All vehicles and containers used for transportation of fish shall be constructed in such a way that they can be easily cleaned and disinfected. Before and after transportation of fish the containers and vehicles shall be cleaned and disinfected.
2. The interior surface of the vehicles shall always be maintained in such a way that they are free from abnormal odours.
3. Vehicles equipped with refrigeration equipment shall be pre-cooled to a temperature of 2 °C or below before loading starts.

Care in Sale/Processing Units

1. Halls for sale or processing of fish shall be situated in areas where water, electricity and skilled labour are readily available.
2. The hall may preferably be situated in the East-West direction so that direct sunlight will not fall in the hall both in the morning and in the evening.
3. The hall and the sections in it shall be planned and designed in such a way that wind does not blow dust either into the hall or on the product.
4. The roof of the building shall be of such a construction that accumulation of dust can be minimised.
5. The roof-wall joint shall be tight so that rodents will not enter the processing hall.
6. The walls of the fish handling hall shall be cemented and polished or fitted with glazed tiles to a height of at least 1.8 meters from the floor to facilitate easy cleaning.
7. The floor-wall joint shall be rounded to facilitate easy cleaning.
8. The floor of the fish handling hall shall have proper slope so that the water on the floor easily runs into the drain.
9. The bottom portion of the drainage channels shall be rounded to prevent accumulation of dirt.
10. The drainage openings shall be closed with grills of proper size in order to prevent entry of rodents.
11. The premises of the fish processing halls shall be kept neat and clean.

12. At the entry to the fish handling area, adequate number of wash basins (foot-operated) may be fitted and the workers shall be instructed to wash and disinfect their hands before entering the hall.
13. Liquid soap, disposable towels, nail brush and chlorine solution (200ppm.) shall always be kept near all the wash basins.
14. All fish handlers may wash their hands from elbow down using soap solution followed by disinfection in water chlorinated to a level of 200 ppm.
15. There shall be a foot-dip filled with a germicide solution at the entry to the hall and the workers shall be instructed to enter the hall only through the foot-dip.
16. The foot-dips shall be sufficiently big in size so that no worker will enter inside without dipping his/her legs.
17. There shall be sufficient ventilation in the fish handling hall to avoid off-odours.
18. There shall be sufficient light in the hall for reasons of both safety and efficient working.
19. Light, bulbs and fixtures suspended over the fish shall be of safety type to prevent contamination in case of breakage.
20. All doors and windows in the fish handling hall shall be fitted with fly-control nets. Air curtains may be fitted, wherever needed, to prevent entry of flies.
21. The door leading to the hall shall be of self-closing type.
22. The hall shall be constructed in such a way that entry of rodents can be prevented.
23. Areas used for storing edible, products shall be separate and distinct from those used for inedible materials.
24. Fish handling areas shall be completely separated from the areas used for residential purpose.
25. There shall be plentiful supply of potable water to the fish handling hall.
26. The water used for washing fish or for ice manufacture shall be chlorinated to a residual level of 10 ppm.
27. Ice shall be stored on raised platforms in separate insulated room.
28. Ice blocks shall not be dragged on the floor.
29. Saw-dust, gunny bag, tarpaulin *etc.* used for covering ice are liable to contaminate the fish with bacteria of various types.
30. Fish shall not be handled on floor.
31. All fish contact surfaces shall be smooth, free from pits and crevices and of non-absorbent type.

32. Before starting and after finishing each day's work all fish contact surfaces shall be cleaned and disinfected.
33. Factory employees with any communicable disease shall not be permitted to handle fish.
34. Workers having unprotected injury/wound on their palms shall be kept away from work.
35. All fish handlers may have to undergo periodic medical examination including stool culture to detect "carriers".
36. Fish handlers shall wear clean uniforms while on duty.
37. Before starting work, all fish-handlers may wash their hands from elbow down using soap followed by disinfection using chlorine of 200 ppm. strength.
38. Spitting, smoking, excessive talk and use of tobacco shall not be permitted in premises where fish is handled.
39. All doors of the processing unit opening to the outside shall be fitted with automatically working air curtains.
40. All processing units should have separate change rooms for gents and ladies. Workers will change their street dress and wear official uniform in the change room. They will also wear gumboots. There should be provision to keep their street dress and street chappal. There should be lockers (to keep the valuable items of workers) and emergency toilets in the change room.
    (a) The primary processing and processing should preferably be integrated.
    (b) On no occasion should there be, back tracking on the flow of the material processed.
    (c) The material should enter the factory through air-curtain fitted chute provided near the raw material receiving platform This platformshould be covered on the top and sides.
    (d) Water from the pocessing tables shall not fall on the floor.
    (e) There should be chill rooms to keep the incoming raw material and the final raw product.
    (f) There should be a separate carton store and the cartons should be kept on racks with covering to protect from dirt and dust.
    (g) There should be an approved system to treat the effluents.
    (h) All activity in the processing unit should be scheduled as outlined under the HACCP system with proper documentation.
    (i) All detergents, disinfectants, antioxidants and additives used in the processing units have to be approved by the competent authority.

(j) The unit should have an approved system for rodent, fly and vermin control.

(k) The water, ice, disinfectants and additives used in the processing unit have to be tested at periodic intervals for the parameters as specified by the competent authority.

41. All processing units should be equipped with an in-house laboratory to monitor plant sanitation and to have an in-process check on the product.

These sanitary/hygienic precautions, if strictly followed, will definitely improve the quality, of fish reaching the consumer.

4.4 Food Safety with Special Reference to HACCP System

Over a period of time, there has been considerable change in the concept of food consumption. The habit of eating outside has increased. Many types of ready-to-eat food items are available in the market and these are purchased and consumed by people without any hesitation. The modern food processing industry is very much sophisticated and technologically advanced. Further, several ingredients are now added to foods as additives, antioxidants, preservatives, emulsifiers, cyoprotectants and colouring materials. There are also problems of pesticide residues, toxic metals, mycotoxins, biotoxins, antibiotic residues and the like. Under these circumstances, the responsibility of the processor has become increasingly complex and hence, there is a global shift from food quality to food safety. This has resulted in the development of a safety-oriented quality system, the Hazard Analysis Critical Control Point (HACCP) system. This is a world-wide systematic and preventive approach that addresses physical, chemical and biological hazards through anticipation and prevention rather than through end product inspection and testing. The HACCP system consists of seven principles, which are implemented in twelve steps. The advantages of the system are highlighted and the current problems in the implementation of the HACCP system are discussed.

4.4.1 The HACCP Concept

HACCP is a world-wide recognized systematic and preventive approach that addresses physical, chemical and biological hazards through anticipation and prevention rather than end -product inspection and testing. The HACCP concept was pioneered in the 1960s by the Pillsbury Company, the United States Army and the National Aeronautics and Space Administration (NASA) as a collaborative development for the production of safe foods for the U.S. Space Programme. In 1985, the U.S. National Academy of Science recommended that the HACCP system can be adopted by the food industry to ensure food safety. Later, the International Commission on Microbiological specifications for Food (ICMSF) and the International Association of Milk, Food and Environmental Sanitations also recommended HACCP for food safety.

4.4.2 Advantages of HACCP

The HACCP food safety management system uses the approach of controlling critical points in food handling to prevent food safety problems. The system which is science based identifies specific hazards and takes measures for their control. HACCP is based on prevention and reduces reliance on end-product inspection and testing. The system can be applied in the food industry from 'farm to fork'. It enhances responsibility and degree of control at the level of food industry. Implementing HACCP does not mean undoing the quality assurance procedures or good manufacturing practices already established by the company; it only requires a revision and appropriate integration with the HACCP system. Further, the HACCP system can aid inspection by the Regulatory Authorities and promote international trade by increasing buyers' confidence. The HACCP system should be capable of accommodating advances in equipment design, process and technology.

4.4.3 Applications of HACCP

To apply HACCP system in any food industry sector, the sector should be operating according to the Codex General Principles of food hygiene and appropriate food safety legislation. Management Commitment is essential for effective implementation of HACCP system. The HACCP system works on seven principles and there are 12 steps in the implementation of HACCP. The logic sequence for the application of HACCP is shown in Figure 4.3.

1. Assemble HACCP Team

The first step in the application of HACCP is to assemble a team having knowledge and expertise to develop an HACCP plan. The Team should be multi-disciplinary and should include plant personnel from production, sanitation, quality assurance laboratory, engineering and inspection. It is essential to assemble the right blend of expertise and experience as the Team will collect and evaluate technical data and identify hazards and critical control points. If needed, advise of an External Consultant can be obtained.

2. Describe Product

A full description of the product should be drawn up such as composition, physical/chemical structure (including a_w, pH *etc.*), packaging, shelf life, storage conditions and method of distribution.

3. Identify Intended Use

Intended use of the product refers to its normal use by end-users or consumers. The HACCP Team must specify where the product will be sold and the target group especially if it happens to be a sensitive group (elderly, infants, immuno suppressed and pregnant women)

4. Construct Flow Diagram

It is easier to identify the potential routes of contamination and to suggest

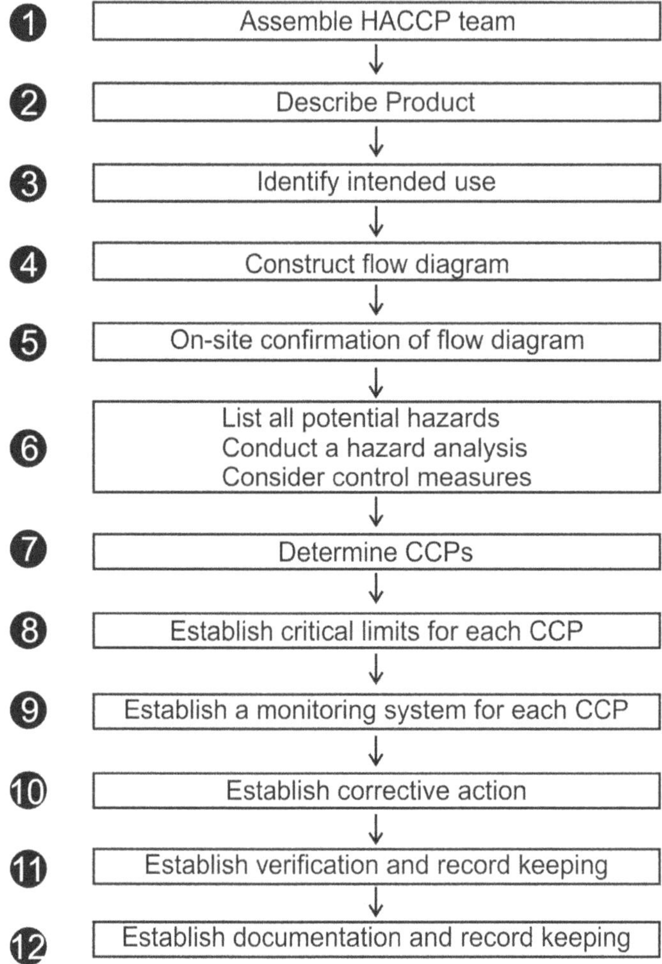

Figure 4.3: Logic Sequence for Application of HACCP.

control measures if there is a flow diagram. The critical control points can also be easily identified. The process flow diagram should identify all the important steps from receipt of raw material to final shipping. Details like ingredients and packaging materials used, time-temperature history flow conditions, product recycle equipment design *etc.* should get reflected in the flow diagram.

5. On-site Confirmation of Flow Diagram

Once the flow diagram has been completed it should be confirmed by on-site inspection for accuracy and completeness. It will also confirm the assumption with respect to the movement of product and employees on the premises.

6. Hazard Analysis

Hazard analysis is the first step of HACCP principle. As is evident from the title, this step is the most important in the HACCP system. Any mistake in hazard analysis will lead to inadequate HACCP plan. Hazard analysis requires technical expertise and scientific background in various aspects of food science and technology for proper identification of all potential hazards. In the HACCP plan, hazard analysis is essential to identity which hazards are of such a nature that their elimination or reduction to acceptable level is essential for production of a safe food. Hazards will vary among firms making the same product because of difference in:

- ☆ Sources of ingredients
- ☆ Formulations
- ☆ Processing equipments
- ☆ Process/preparation methods
- ☆ Duration of processes
- ☆ Storage conditions
- ☆ Experience/Knowledge of personnel
- ☆ Attitude of food-handlers

Hazard analysis is to be conducted for all existing and new products. Whenever there is a change in raw materials, product formulations, processing procedures, packaging, distribution and use of the product, the original hazard analysis is to be reviewed. In conducting hazard analysis the following points are to be considered.

- ☆ The lively occurrence of the hazard, its severity and adverse effects on health. survival and multiplication of microorganisms.
- ☆ Production and viability of toxins.
- ☆ Production of chemicals and condition leading to their production.

Hazards are classified as Physical, Chemical and Biological. The hazards are listed.

I. Physical Hazards

Physical hazards include any potentially harmful extraneous matter not normally found in food. The extraneous matter found in fish products can be divided or classified as:

1. Non-food safety hazards (*e.g.* filth)
2. Food safety hazards (*e.g.* glass, metal, wood, bones, stones, hard plastic). The adverse health effect of physical hazards may be choking, injury incl. laceration and perforation of tissues in the mouth, throat, stomach or intestines. Broken teeth anddamage to gums may also be the result. The FDA Health Hazard Board has found that foreign objects that are less than 7 mm maximum dimensions rarely cause trauma or serious injury

except in special riskgroups such as infants, elderly or surgery patients (FDA, 1998)Although physical hazards rarely cause serious injury, they are among the most commonly reported consumer complaints, because the injury occurs immediately or soon after eating, and the source of the hazard is often easy to identify.

Control measures for physical hazards can include:

1. For metal inclusions: periodically checking all equipment for damage or missing parts passing the product through metal detection or separation equipment.
2. For non-metallic objects: passing the product through an X-ray detector.

II. Chemical Hazards

Naturally occurring	Added chemicals	Form Packaging material
Chemicals	Pesticides	Vinyl chloride
Allergens	Fertilizers	Printing inks
Aflotoxins	Antibiotics	Adhesives
Mushroom toxins	Growth hormones	Lead
PSP	Lead	Tin
DSP	Zinc	
ASP	Cadmium	
Ciguatoxin	Food additives	
Contaminants		
☆ Lubricants		
☆ Detergents		
☆ Sanitizers		
☆ Paints		
☆ Refrigerants		

The marine environment particularly, the coastal seas and eustarine zones are subjected to a high degree of contamination from many anthropogenic compounds and industrial pollutants, causing a direct threat to aquatic life or an indirect hazard to fish eaters. Chemical contaminants comprise mainly of industrial and environmental chemicals such as heavy metals, chlorinated pesticides, persistent organo-halogen compounds polychlorinated biphenyls (PCBs), persistent pollutants (POPs) and Organo chlorine pesticides. OCPs have been recorded in cetaceans throughout the world. High concentration of PCBs have been associated with reproductive failure in common seals and implicated in recent epizootics involving common seals in the North sea and stripped dolphin in the Mediterranean sea. Similarly elevated levels of toxic metals like Cd, Hg, Pb, As, Cu, Zn, Ni *etc.* have been observed in various fishery products, particularly molluscan shellfish, from different parts of the world. Degradation of the quality of the environment with chemical contaminants has caused concern, the world over.

The Codex Alimentarius Commission of the U.N. aims at global consumers health and economic interests and fair practices in the trade of food. As a result of stringent quality requirements and standards, the seafood exporting as well as importing countries of the world has taken serious view about the safety of the food products they market. With the implementation of EU directive, (EU 91/493/EEC) and the US regulations of 1997, it has become mandatory to monitor the levels of various chemical contaminants in seafood. Chemical contaminants can occur at any stage in food production, processing and marketing. Regulatory limits are set for some of these contaminants.

The important groups of chemical contaminants in seafood causing public health hazard are described below. The sources of such contaminants could be natural processes, urban seawage discharge or industrial or agricultural waste discharge. Some chemicals (food additives) are intentionally added to enhance the shelflife or the quality and acceptability of the product or to protect the product from further decomposition.

A) Toxic Metals in Seafood

Catastrophical incidence of heavy metal poisoning by eating contaminated fish/shellfish were reported from various parts of the world. Some of the water bodies or eustrine systems are heavily polluted by metals, *viz.*, Derwent estuary, Tasmania, Sor Fijord (Norway), Rathrongut estuary (UK) and Rio Tintu estuary (Spain). The impaired quality and safety of seafood may lead to unacceptability and rejection in the overseas markets. The incidents of heavy metal poisoning started with the mercury poisoning episode of Minamata reigon in Japan. Disease of the dancing eat implicated by mercury poisoning, 'Itai-itai' disease caused by Cd poisoning and the black foot disease caused by As poisoning (As poisoning caused by drinking contaminated water) observed recently in Bangladesh *etc.* are examples. A ten nation study disclosed that Pb level in human blood varied markedly from region to region. Elements of most concern are cumulative poisons that include Hg, Cd, Pb, Cr, Se and As.

i. Mercury and Methyl Mercury in Fish

When exposed to Hg containing diet or water, fish, shellfish *etc.* generally accumulate the metal throughout their life. Food chain enrichment of Hg is taking place in the aquatic environment. Eventually, man consuming the fish, inevitably suffers from the results of this enrichment taking place at each trophic level. In general, the level of Hg from unpolluted area is in the range of 0.01 – 0.3 ppm and in fish from polluted water the level goes up to 2.5 ppm. Fish from the polluted Minamata region had total mercury content of about 50 ppm. High concentration of mercury were found in tuna and sword fish (Table 4.1) particularly in large size. However, the value remained unaltered over a century. The mean mercury content in tuna and sword fish were 0.91 and 3.1 ppm respectively. Most small or medium sized tuna contained levels that fall far below the limit. The mercury content (Hg) found in Indian molluscan products were quite low and none of the samples had a value exceeding 0.5 ppm (the highest level was found in oysters <0.3 ppm and Hg in Indian fishery products ranged between 0.03 to 0.336 ppm in shellfish).

Table 4.1: Mercury Content in Fish

Fish	*Range of Values, ppm*	*Mean Hg, ppm*
Tuna fish, (*Museum* sp.)	0.03–1.51	0.95 dry wt. Basis
Tuna fish, (Recent sp.)	0.44–1.53	0.91 dry wt. basis
Sword fish (Recent sp.)	0.94–5.08	3.1 ± 1.5 dry wt. Basis
Squid, whole	0.03–0.09	0.066
Cultletish, whole	0.04–0.10	
Mussel, *Perna viridis*	0.02–0.049	0.038

ii. Cadmium

Contamination of the aquatic environment with Cd is less widespread. However, in recent years higher levels of Cd was observed in marine products from certain regions. Higher levels of Cd in cephalopods have been reported from different parts of the world. Around 50 per cent of squid and cuttlefish imported to Italy and Yogoslav had exceeded the tolerance limit for Cd. (1 mg kg^{-1}). Falandyz (1989) found higher levels of Cd (2.9-10-1 mg kg^{-1} wet wt.) in the edible parts of squid, *Loligo patagonica*. Raw whole squid contained on an average 4.0 mg. Cd Kg^{-1} (Falandyz 1991). The cephalopod, export from India also suffered a set back during 1980s due to Cd contamination.

Sources of Cd in the environment are from zinc smelting, electroplating industry, pigment plants and producers of alloys and batteries. Significant quantities of Cd will reach the agriculture soil from phosphatic fertilizers. Coal and oil combustion, cement production and incinerations of waste product would release huge quantities of Cd in the atmosphere.

The first reported Cd poisoning occurred in Japan during 1947. The disease which is, of a rheumatic nature was subsequently known as 'itai-itai' disease. Toxicological symptoms include renal dysfunction and osteomalacia.Higher levels of Cd have been observed in whole cephalopods, oysters and similar species. Cadmium accumulates, mainly in the liver and kidney of cephalopods. A detailed account of metal distribution (Cd etc) in cephalopods have been presented in the report on cephalopod study. (Lakshmanan 2001). Tolerance limit in seafoods varied between nation and it is in the range of 1-3 ppm.

iii. Lead

Lead is another cumulative poison. Lead poisoning causes chronic kidney infection. The toxicity of Pb to marine organisms is not known precisely. Shellfish taken from inshore waters were found to contain high levels of lead. Mussels from the Derwent estuary (Tasmania) were found to be unfit for human consumption because of their high lead content (Hg, Zn, Cd and Pb). Lead content varied from 16-532 ppm (dry wt) in mussels from the estuary. This can cause risk to health, if contaminated fish is eaten in excess. Chronic lead poisoning in humans from contaminated food or water is common.

Lead Toxicity

Lead toxicity is generally known as plumbism and is generally taking place through water pipes. A ten nation study revealed that lead levels in human blood varied markedly from one country to another.

iv. Arsenic

As is another cumulative poison and in areas of high As concentration, shellfish contain as much as 100 ppm. 100 mg. of As may cause poisoning in man and 130 mg has proved fatal. Arsenic along the food chain within a defined polluted area is as follows:

		Range in ppm
Mussels	–	14.1–16.7
Prawns	–	62.9–80.2
Fish species (assorted)	–	43.4–188 ppm (dry wt. basis)

Irrespective of environmental pollution, many marine organisms contain appreciable amounts of As in their tissue. These authors reported that carnivorous gastropods are characterized by high levels of As (16.8 to 67.9 ppm) in soft parts especially in muscle. The highest As content was found in an American Place (188 ppm).

Arsenic Poisoning

Black foot disease has been reported recently in many people from Bangladesh due to high As content in drinking water. A recent study from Bangladesh revealed As poisoning in humans implicated by high As levels in drinking water. The levels of As in fishery products are not at alarming levels and are far below the limit of FDA guidance level. As content in cuttlefish and clam meat from India is around 15 ppm.

v. Selenium

Food chain enrichment of Se is noted with high building up in the liver of fish. Large specimens of tuna and sword fish contain relatively high levels of Se. Black marlin from Australia contained 2.2 ppm Se in the body. The element is highly toxic to man, with safe level in food considered to be not more than 3 ppm. Se and its compounds cause serious injury to the kidneys, liver and heart.

C. Food Additives

Food additives, including colour additives are used in many fish and fishery products. Food additives have been defined by FDA as chemical that when added to food tends to prevent or retard deterioration and does not include common salt, sugars, vinegars, spices or oil extracted from spices. Before using a food additive the processor should become familiar with the applicable legal allergic type reactions if not properly used. The E.U. directive No.95/2/EC on Food additives says that only additives which satisfy the requirements laid down by the scientific committee for food may be used in food stuffs. The directive has listed various food additives and some of these additives are dealt in this article.

i. Sulphiting Agents

Comprise mainly sodium meta-bisulphite, sodium sulphite, pot. metabisulphite, pot. Sulphite and sulphur dioxide. They are included in the GRAS list and have been universally used. However, since it implicates allergic reactions in some individuals its use has been restricted. Thus, in raw shrimp the MRL prescribed is only 100 ppm.

ii. Benzoic Acid and Sodium Benzoates

Are often used as antimicrobial agents in food preservation. Their excessive intake and may cause necrosis of hepatocytes.

iii. Nitrates and Nitrites

Are also used as preservatives, but excess nitrate induces methmyoglobin – At low doses nitrite is not toxic, but may enhance toxicity or carcinogenicity of other compounds (hyperplasia induced by phenolic compounds).

iv. Antioxidants

Antioxidants are known to have carcinogenic effects. *e.g.*, BHT enhances thyroid hyperplasia and BHA (Butylated hydroxy anisole) can act as promoters of bladder carcinogenesis.

Maximum Residual Limits (MRLs) for pesticides, heavy metals and antibiotics and other pharmacologically active substances in fish and fishery products shall meet the requirement as given in Table 4.2.

D. Antibiotic Residues in Fishery Products

Antibiotics (tetracycline group) are widely used in shrimp culture to control disease. The usual practice is to withdraw the drug well in advance, before harvest, to enable the animals to clear off the antibiotic residue. However, cultured shrimps are sometimes contaminated with the residue. During 1992-94, many consignments of shrimp, exported by Thailand were rejected by Japan owing to their antibiotic residues. Japanese Food sanitation law prohibits antibiotic residues in food. From Nov. 1994 onwards US-FDA continued testing of antibiotic residues in cultured salmon and shrimp.

Table 4.2: Maximum Permissible Residue Level of Heavy Metals, Pesticides and Antibiotics (in ppm)

Heavy Metals		*Pesticides*		*Antibiotics*	
Mercury	1.0	BHC	0.3	Tetracycline	0.1
Cadmium	3.0	Aldrin	0.3	Oxytetracycline	0.1
Arsenic	75.0	Dieldrin	0.3	Trimethoprim	0.05
Lead	1.5	Endrin	0.3	Oxolinic acid	0.3
Tin	250.0	DDT	5.0	Nalidixic acid	Nil
Nickel	80.0			Sulphamethoxazole	Nil
Chromium	12.0			Chloramphenicol	Nil
				Furazolidone	Nil
				Neomycin	Nil

Antibiotics such as penicillin and sulfonamides trigger allergic type reactions in humans. Widespread use of antibiotics develop antibiotic resistence in pathogens.*e.g.* To quinolines in *Campylobacter* spp. and to chloramphenicol in *Yersinia, tetracycline* and *ampicillin in Salmonella*. Consequent on the detection of nitrofuran residues by EU in Indian shrimps all farmers and their Associations have been warned not to use any form of nitrofurans in aquafarms.

a. Prohibited Antibiotics

i. All Nitrofurans including
   - ☆ Furaltadone
   - ☆ Furazolidone
   - ☆ Furylfuramide
   - ☆ Nifuratel
   - ☆ Nifuroxime
   - ☆ Nifurprazine
   - ☆ Nitrofurantoin
   - ☆ Nitrofurazone

ii. Chloramphenicol

iii. Neomycin

iv Nalidixic acid

v. Sulphamethoxazole

vi. Aristolochia spp and preparations thereof

vii. Chloroform

viii. Chlorpromazine

ix. Colchicine

x. Dapsone

xi. Dimetridazole
xii. Metronidazole
xxiii. Ronidazole
xiv. Ipronidazole
xv. Other nitroimidazoles
xvi. Clenbuterol
xvii. Diethylstibestrol (DES)
xviii. Sulfonamide drugs
(except approved sulfadimethoxine, sulfabromomethazine and Sulfaethoxypyridazine)
xix. Fluoroquinolones
xx. Glycopeptides

E. Food Allergens

Food allergens are proteins and constrained to molecular dimensions 10-70 KD in size. These values reflect the usual method of allergen characterization (Sodium dodecayl sulphate polyaerilamide electrophoresis).

Most food allergens are glycoproteins possessing acidic isoelectric points. Most allergenic foods contain multiple allergens. For example, egg white is a complex mixture of 20 proteins, only 5-6 are allergenic. Common food allergens that have been fully or partly purified are cod fish, cows milk, peanuts, soyabeans and shrimp. Most common allergenic food proteins are heat stable and resistant to proteolytic processes. They are resistant to pH extremes, however, are susceptible to digestion, cooking and proteolytic processes.

F. Biotoxins in Seafood

Marine biotoxins are responsible for a number of seafood borne diseases. Two main types of poisoning by marine animals are recognized. Biotoxins are severe hazards in seafood. Biotoxins such as PSP, DSP, NSP *etc.* are fatal and are caused due to periodic marine dinoflagellate blooms that intoxicate sedentary animals like mussels, clams *etc.*

Table 4.3: Biotoxins is Seafood

Sl.No.	*Toxin*	*Where/When Produced*	*Animals/Organ Involved*
1.	Tetrado toxin (puffer fish poisoning)	In fish	Puffer fish, mostly ovaries, liver and intestine
2.	Ciguatera	Marine algae (toxic dinoflagellates) *e.g. Gambierdiscus toxicus*	>400 tropical and sub tropical fishes are involved (reef fish) *e.g.* red snapper grouper, moray eel

Sl.No.	Toxin	Where/When Produced	Animals/Organ Involved
3.	Paralytic shellfish poisoning (PSP)	Marine algae (dinoflagellate blooms) >10^6 cells/litre are required, *e.g. Alexandrium catenella, Gonyaulax tamorensis*	Mussels, Clams, Oysters, Scallops etc. Highly toxic (1g of mussel meat can kill 5 people)
4.	Diarrhetic shellfish poisoning (DSP)	Accumulated by shell fish from dinoflagellate (*e.g. Dinophysis fortii*	Mussels and Clam - Toxin is okadaic acid and its derivative
5.	NSP	Brevitoxin, marine dinoflagellate *Karenia brevis* (formerly known as *Gymnodinium breve* and *Phychodiscus brevis*)	Mussels and Clam
6..	Polytoxin Scombrotoxin (Histamine)	*Palythoa* spp. Scombridae and clupeidae species by *Proteus morgani, Morgenella morgani,* Hafni etc.*e.g.*- tunas, bonitos mackerel, blue fish, dolphin fish *etc.*	Corals Scombroid fishes

Permissible Limits in Seafoods

Histamine : <50mg/100g muscle

PSP : <80 µg/100g muscle

DSP : <20 µg/100g muscle

III. Biological Hazards

Bacteria (Spore forming)	Non-spore Forming	Virus, Protozoan and Parasites
Clostridium botulinum	Salmonella	Hepalitis A and E
Clostridium perfringens	Shigella	Rotavirus
Bacillus ccrcus	Staph.aurcus	Renia solium
	V. parahaemolyticus	*Taenia saginata*
	V. vulnificus	*Entamoclria hystolitica*
	V. cholerae	
	L. monocytogenes	

After hazard analysis is completed, the Team must consider what control measures can be applied for the control of each hazard. Control measures are actions that can be used to prevent or eliminate a food safety hazard or to reduce it to an acceptable level. More than one measure may be required to control a specific hazard or more than one hazard may be controlled by a specific measure.

7. Determine Critical Control Points

This is the second principle of HACCP. It is a step at which control can be applied and it is essential to prevent or eliminate a food safety hazard or reduce it to an acceptable level. If a hazard has been identified at a step where control is necessary for safety and if no control measure exists at that step or any other, then the product or process should be modified at that step, at an earlier step or at a later step to include a control measure. Determination of CCPs in the HACCP system can be facilitated by the application of a decision tree. In a processing system, CCP is the point where hazards need either to be prevented, eliminated or reduced to acceptable levels. For example, time-temperature monitoring is a CCP in a cooking line, where a particular temperature for a particular period will destroy all microorganisms. Similarly, refrigeration to prevent hazardous microorganisms, pH measurement to prevent bacterial growth and metal detector on the packaging line can be considered as Critical Control Points.

8. Establish Critical Limits for each CCPs

Critical limits are criteria that separate acceptability from unacceptability. In other words, critical limits are boundaries used to judge whether a safe product is being produced. Critical limits can be set for parameters like temperature, time, pH, a_w, moisture level *etc.* Critical limits should meet government regulations, company's standards and should be supported by scientific data. Sources of information on critical limits can be obtained from:

- ✰ Scientific publications
- ✰ Government regulations
- ✰ Expert
- ✰ Experimental studies

If the required information on critical limits is not available, a conservative value should be selected. The reference material used should be recorded and this should become part of the support documents for the HACCP plan.

9. Establish Monitoring System for each CCPs

Monitoring is the act of conducting a planned sequence of observations of control parameters to assess whether the CCP is under control. Monitoring procedures should be efficient enough to detect loss of control at the CCP. It is important to specify how, when and by whom monitoring is to be performed. There are several ways to monitor the critical limits at the CCPs. It can be done continuously or on batch basis. Continuous monitoring is preferred as it is possible to identify shifts around the critical limits. When monitoring is not continuous, the frequency of monitoring should be specified. A further consideration in considering a monitoring system is the time taken for monitoring. Monitoring procedures should be rapid. Chemical measurements and visual observations are preferred to microbiological methods. pH, a_w, moisture, time and temperature are the usual parameters tested. It is essential that all the monitoring equipments are calibrated for accuracy.

10. Establish Corrective Action

Corrective action is the action to be taken when the results of monitoring at the CCP indicate a loss of control. Loss of control will result in a deviation from the critical limits for a CCP. Corrective action should then be taken following any deviation to ensure safety. If the corrective action does not address the root cause of the deviation, then the deviation could recur. Corrective action steps include:

- ☆ Investigation to determine the cause of deviation
- ☆ Effective measure to prevent recurrence in future
- ☆ Verification of the effectiveness of the corrective action taken

11. Establish Verification Procedures

Verification procedures are necessary to assess the effectiveness of the HACCP plan. Periodic verification helps to improve the HACCP plan by exposing weaknesses in the system and eliminating unnecessary or ineffective control measures. Verification activities include:

- ☆ HACCP plan verification
- ☆ HACCP system audit
- ☆ Equipment calibration
- ☆ Targetted sample collection and testing

12. Establish Documentation and Record Keeping

Record shows process history, monitoring, deviations and corrective action including disposition of product that occurred at the identified CCP. Records may be in the form of chart, written record and computerized record. Four types of records are to be maintained.

- ☆ Support documents for developing HACCP plan
- ☆ Records generated by the HACCP system
- ☆ Documentation of methods and procedures
- ☆ Records of employee training programmes

HACCP Study Tips

- ☆ Use disciplined approach
- ☆ Don't rush
- ☆ Don't make assumptions
- ☆ Set deadlines for comments
- ☆ Challenge beliefs
- ☆ Keep accurate records

- Discussion-hierarchical
- Team Leader should moderate, not dominate

Requirements for a Successful HACCP System

- Management commitment
- Plant design as per GMP
- Insect and pest control
- Hygiene and sanitation
- Trained Personnel

Current Problems

Lack of:

- Verification
- Evaluation
- Validation
- Review
- Audit
- Not applied from "farm to fork."

HACCP is not a sophisticated system requiring high technology and highly educated staff. Simple tests like sensory methods. time-temperature evaluation. pH determination *etc.* are employed in this system. The system can be run by the technical staff of the industry after they get adequately trained under an expert.

4.5 Strategies for Prevention of Food Borne Diseases

1. To start food surveillance in order to trace origin of contamination
2. To create consumer awareness on food quality and hazards associated with contaminated and adulterated food stuffs
3. To improve infrastructure of food quality control systems in the country
4. To promote coordination of all departments and agencies implementing quality control
5. Revise national standards on par with international standards
6. Introduce legal and institutional facilities and support for quality control agencies in the country
7. To promote and upgrade manufacturing standards of food processing factories.

   Advise them to introduce Good Manufacturing Practices (GMP) and Hazard Analysis Critical Control Points (HACCP) in their plants.
8. Those who violate food laws must be punished without delay
9. Develop an integrated information network through out the country for monitoring and dissemination of information to public

4.6 Food Safety Management System

Food safety is related to the presence of food-borne hazards in food at the point of consumption (intake by the consumer). As the introduction of food safety hazards can occur at any stage of the food chain, adequate control throughout the food chain is essential. Thus, food safety is ensured through the combined efforts of allthe parties participating in the food chain. Organizations within the food chain range from feed producers and primary producers through food manufacturers, transport and storage operators and subcontractors to retail and food service outlets (togetherwith inter-related organizations such as producers of equipment, packaging material, cleaning agents, additives and ingredients). Service providers are also included.

This International Standard specifies the requirements for a food safety management system that combines the following generally recognized key elements to ensure food safety along the food chain, up to the point of final consumption:

- Interactive communication;
- System management;
- Prerequisite programmes;
- HACCP principles.

Communication along the food chain is essential to ensure that all relevant food safety hazards are identified and adequately controlled at each step within the food chain. This implies communication between organizations both upstream and downstream in the food chain. Communication with customers and suppliers about identified hazards and control measures will assist in clarifying customer and supplier requirements (*e.g.*with regard to the feasibility and need for these requirements and their impact on the end product).

Recognition of the organization's role and position within the food chain is essential to ensure effective interactive communication throughout the chain in order to deliver safe food products to the final consumer. The most effective food safety systems are established, operated and updated within the framework of a structured management system and incorporated into the overall management activities of the organization.

This provides maximum benefit for the organization and interested parties. This International Standard has been aligned with ISO 9001 in order to enhance the compatibility of the two standards. This International Standard can be applied independently of other management system standards. Its implementation can be aligned or integrated with existing related management system requirements, while organizations may utilize existing management system(s) to establish a food safety management system that complies with the requirements of this International Standard.

This International Standard integrates the principles of the Hazard Analysis and Critical Control Point (HACCP) system and application steps developed by the Codex Alimentarius Commission. By means of auditable requirements, it combines the HACCP plan with prerequisite programmes (PRPs). Hazard analysis is the key toan effective food safety management system, since conducting a hazard analysis assists in organizing the knowledge required to establish an effective combination of control measures. This International Standard requires that all hazards that may be reasonably expected to occur in the food chain, including hazards that maybe associated with the type of process and facilities used, are identified and assessed. Thus it provides the means to determine and document why certain identified hazards need to be controlled by a particular organization and why others need not.

During hazard analysis, the organization determines the strategy to be used to ensure hazard control by combining the PRP(s), operational PRP(s) and the HACCP plan.

This International Standard specifies requirements for a food safety management system where an organization in the food chain needs to demonstrate its ability to control food safety hazards in order to ensure that food is safe at the time of human consumption. It is applicable to all organizations, regardless of size, which are involved in any aspect of the food chain and want to implement systems that consistently provide safe products. The means of meeting any requirements of this International Standard can be accomplished through the use of internal and/or external resources.

This International Standard specifies requirements to enable an organization:

a. To plan, implement, operate, maintain and update a food safety management system aimed at providingproducts that, according to their intended use, are safe for the consumer,
b. To demonstrate compliance with applicable statutory and regulatory food safety requirements,
c. To evaluate and assess customer requirements and demonstrate conformity with those mutually agreed customer requirements that relate to food safety, in order to enhance customer satisfaction,
d. To effectively communicate food safety issues to their suppliers, customers and relevant interested parties in the food chain,
e. To ensure that the organization conforms to its stated food safety policy,
f. To demonstrate such conformity to relevant interested parties, and
g. To seek certification or registration of its food safety management system by an external organization, or make a self-assessment or self-declaration of conformity to this International Standard.

All requirements of this International Standard are generic and are intended to be applicable to all organizations in the food chain regardless of size and complexity.

This includes organizations directly or indirectly involved in one or more steps of the food chain. Organizations that are directly involved include, but are not limited to, feedproducers, harvesters, farmers, producers of ingredients, food manufacturers, retailers, food services, catering services, organizations providing cleaning and sanitation services, transportation, storage and distribution services. Other organizations that are indirectly involved include, but are not limited to, suppliers of equipment,cleaning and sanitizing agents, packaging material, and other food contact materials.

This International Standard allows an organization, such as a small and/or less developed organization (*e.g.* a small farm, a small packer-distributor, a small retail or food service outlet), to implement an externally developed combination of control measures.

4.6.1 General Requirements

The organization shall establish, document, implement and maintain an effective food safety management system and update it when necessary in accordance with the requirements of this International Standard.

The organization shall define the scope of the food safety management system. The scope shall specify the products or product categories, processes and production sites that are addressed by the food safetymanagement system.

The organization shall:

a. ensure that food safety hazards that may be reasonably expected to occur in relation to products within the scope of the system are identified, evaluated and controlled in such a manner that the products of the organization do not, directly or indirectly, harm the consumer,
b. communicate appropriate information throughout the food chain regarding safety issues related to its products,
c. communicate information concerning development, implementation and updating of the food safety management system throughout the organization, to the extent necessary to ensure the food safety required by this International Standard, and
d. evaluate periodically, and update when necessary, the food safety management system to ensure that the system reflects the organization's activities and incorporates the most recent information on the food safety hazards subject to control.

Where an organization chooses to outsource any process that may affect end product conformity, the organization shall ensure control over such processes. Control of such outsourced processes shall be identified and documented within the food safety management system.

4.6.2 Documentation Requirements

i. General

The food safety management system documentation shall include:

a. documented statements of a food safety policy and related objectives (see 5.2),
b. documented procedures and records required by this International Standard, and
c. documents needed by the organization to ensure the effective development, implementation and updating of the food safety management system.

ii. Control of Documents

Documents required by the food safety management system shall be controlled. The controls shall ensure that all proposed changes are reviewed prior to implementation to determine their effects on food safety and their impact on the food safety management system.

A documented procedure shall be established to define the controls needed:

a. to approve documents for adequacy prior to issue,
b. to review and update documents as necessary, and re-approve documents,
c. to ensure that changes and the current revision status of documents are identified,
d. to ensure that relevant versions of applicable documents are available at points of use,
e. to ensure that documents remain legible and readily identifiable,
f. to ensure that relevant documents of external origin are identified and their distribution controlled, and
g. to prevent the unintended use of obsolete documents, and to ensure that they are suitably identified as such if they are retained for any purpose.

iii. Control of Records

Records shall be established and maintained to provide evidence of conformity to requirements and evidence of the effective operation of the food safety management system. Records shall remain legible, readily identifiable and retrievable. A documented procedure shall be established to define the controls needed for the identification, storage, protection, retrieval, retention time and disposition of records.

4.7 Quality Standards

Food standards have been introduced on a national/international basis to protect the consumers health and to ensure fair practices in food trade. The formulation of standards for fish and fish products became necessary to attain a minimum standard of cleanliness and hygiene in fish handling, processing and marketing. The exporting country or company should be aware of the quality requirements of the buying nation. Standards are intended to guide and promote export or import of fishery products between countries. Since Governments of all countries are responsible for public health problems arising from the consumption of fish products, they enforce certain food laws and introduce standards. These standards fall into two main categories.

4.7.1 Safety Standards

Safety standards are formulated to protect the consumer against food that are damaging to health. This ensures that reasonable standards of hygiene are practiced so that fish are free of pathogens and that use of food additives are controlled and contaminants are prevented.

4.7.2 Composition Standards

These standards protect the consumer against fraud by ensuring that food is unadulterated, pure and of good quality. Packages should contain the correct description, labeling, weights *etc.* Examples are fish pastes and fish fingers where the composition is printed on the label.

The main concern of the food laws is therefore the safety identification, quality labeling and advertising, both to inform and to protect the consumer and sustain a fair basis for honest trading. In addition to these food laws, various national and international standards and codes of practices exist in order to place good quality and safe products in the market. The different standards in operation are:

1. National Standards (BIS, BS, US FDA *etc.*)
2. International standards (FAO, Codex Alimentarius)
3. Company specific standards

4.8 National Standards

Many fish producing countries have their own standards and codes of practice for fishery products. The Bureau of Indian Standards, BIS (former Indian Standards Institution ISI); British Standards, BS; United States Food and Drug Administration, US FDA *etc.* have brought out standards for their fish and fishery products. They govern the quality and standards of products (fish/fishery) for local consumption as well as those for export and import. In U.K., the White Fish Authority and Herring Industry Board have published code, detailed minimum standards for a range of chilled and frozen products. Codes of practices are also available relating to hygiene in the retail industry and in handling and transportation of fish. These are complemented by the British Standards Institutions' "Recommendations on cleaning in the fish industry" (BS 4259,1968).

The Food and Drug Administration (FDA) is engaged in the formulation of processing standards, in the inspection of imported products and in the public health surveillance of processing establishments. The National Marine Fisheries Service (NMFS), US, sets 'grade standards', for fish products using a three-grade system for 15 or more major frozen products. Products meeting the standards after inspection are allowed to bear three descriptions (Grade A, B or Substandard) depending upon quality. In the area of processing standards, the FDA has drawn up guidelines on Good Manufacturing Practice for food in general and two fish products in particular. Fish and fishery products exported to US must meet the requirements of FDA standards. Mandatory inspection of chilled and frozen fish landed at Japanese ports from fishing vessels is carried out by highly trained officials employed by the Food Inspection Service. The aspects included are :

i. Checking for spoilage or contamination
ii. Bacterial testing of raw shellfish
iii. Ensuring that edible fish containing poisonous organisms are identified and segregated
iv. Ensuring that adequate sanitary conditions prevail.

Detailed mandatory standards are also in force in conjunction with compulsory inspection of canned and frozen products.

The most important considerations for food quality standards are safety of the consumer and quality identification of the products. A minimum standard of cleanliness and hygiene in handling of fish products are required to ensure the health of consumers. For this the exporting country should be aware of the quality requirements emphasized by the importing country. The safety regulations are intended to guide and promote export or import of fishery products that are recognized nationally and internationally.

Two categories of standards are introduced to enforce laws and regulations regarding food safety. The foremost are the safety standards which are formulated to protect the consumer against food that are injurious to health. The safety standards ensure hygiene practices so that the fish are free of pathogens. The second one is composition standards that are intended to protect the consumer against fraud and adulterated food ensuring pure and good quality food. For this the packages are marked for proper description such as composition, grade, weights, brand *etc.*

A number of national and international standards and codes of practices exist in the trade all over the world. Important national standards are BIS, BS, USFDA *etc.* The well known international standards are ISO standards, and FAO codex Alimentarius. Fish product standards are predominantly voluntary, and are meant to refer to a reasonably complete and applied fish product specifications that have been agreed nationally or internationally (Connell, 1980). The term regulatory is applied to standards used in conjuction with official inspection or certification schemes.

A number of National Standards and Codes of Practice are available relating to fish and fishery products. British Standards Institution is the foremost institute in this respect and they have provided, quality standards for fish and fishery products for domestic and also for export and import. The Bureau of Indian Standards, BIS, is the Indian counterpart which has brought out standards for fish and fishery products. Likewise Food and Drug Administration of US is concerned with formulation of processing standards in the inspection of products and in the public health surveillance of the processing establishments. It has also drawn up guidelines on good manufacturing practice for food in general and also fish products. The following regulations issued by FDA are an important part of food and Drug Law.

a. Current Good manufacturing Practice Regulations.
b. New Drug regulations
c. FDA food standards

Principal Requirements of Food Law

1. Health safe guards
2. Economic safe guards
3. Labeling requirements
4. Sanitation requirements and personnel hygiene requirements
5. Current good manufacturing practice regulations
6. Plant construction design and layout
7. Defect action levels
8. Food additives
9. Warehousing
10. Food standards
11. HACCP implementation

4.8.1 Bureau of Indian Standards

The National Standards Organization in India is the Bureau of Indian Standards (BIS). Its main aim is to prepare standards on national basis and promote their adoption. The Bureau has brought out over fifty standards for various fishery products. These standards prescribe detailed requirements of processing, packaging and methods of analysis for evaluation of quality of the product. A list of Indian Standards for fish and fishery products is given in Tables 4.4–4.6. These aspects cover scope of standards, terminology, grades, preparation of material, quality requirements, packaging, marking, sampling and testing physical, sensory, microbiological and chemical aspects.

Standards are available for pomfrets, mackerel, seer fish *etc.* Formulation of standards have been felt necessary to make available fresh fish of desired quality and raw material for freezing and canning purpose based on characteristic parameters like fresh appearance, colour, odour, bright eyes and skin, bright red gill, firm flesh and internal odours.

i. Fresh Fish

Under fresh fish, standards are available for pomfrets, mackerel, threadfin, seer fish *etc.* The formulation of the standards has been felt necessary with a view to make available fresh fish of desired quality. It should be handled and transported under sanitary conditions, washed in water containing 5-10 ppm chlorine, precooled and iced. The material shall be clean, wholesome and fresh. It should have characteristic colour, odour, bright eyes, bright red gill, firm flesh *etc.* The material may also satisfy the microbiological standards.

ii. Frozen Fish

Indian Standards are available for frozen products like shrimp, lobster tails, crabmeat, cuttlefish, squid, pomfret, threadfin, mackerel, seer fish *etc.* The main quality problems of the frozen products are dehydration, under weight, size grade, discoloration, decomposition and microbiological requirements. On thawing, the product should be clean, sound, intact, undamaged and free from defects. Deteriorations such as dehydration, rancidity and adverse changes in the texture shall not be present. The products shall be free from foreign matter.

iii. Dried and Cured Products

Under this, standards are available for dried prawns, dried white baits, dried and laminated Bombay duck, dry salted products like mackerel, seer fish, shark, tuna, threadfin, jewfish, catfish, horse mackerel *etc.* The main quality requirements are:

- Material shall have characteristic dry-salted fish odour and shall not show red or pink discoloration.
- Shall be free from off-odour indicative of spoilage.
- Freedom from foreign matter.
- Freedom from excessive sand and salt.
- Freedom from insects and mite infestation and from visible fungal growth.
- The content of moisture, salt and ash are also prescribed.
- The curing period is also specified. The fish while drying shall be protected against contamination from dirt, sand, flies and insects.

Standards are also available for a variety of canned products (Table 4.5), fish meal, shark liver oil and sardine oil. In addition to these, standards are also prepared for code of practices for different operational practices like:

- Code for hygienic conditions for fish industry (Pre-processing and Processing units),
- Recommendations for maintenance of cleanliness in fish industry,
- Basic requirements for fresh fish stalls,
- Requirements for a fish market,

- Procedure for checking temperature of quick frozen foods and
- Specification for master cartons for export of fishery products
- Methods of tests to achieve the various quality standards

4.9 Export Inspection Council of India

The Export Inspection Council (EIC) was set up by the Government of India under Section 3 of the Export (Quality Control and Inspection) Act, 1963 (22 of 1963), in order to ensure sound development of export trade of India through Quality Control and Inspection and for matters connected thereof. EIC is an advisory body to the Central Government, which is empowered under the Act to:

- Notify commodities which will be subject to quality control and/or inspection prior to export,
- Establish standards of quality for such notified commodities, and
- Specify the type of quality control and/or inspection to be applied to such commodities.

Besides its advisory role, the Export Inspection Council, also exercises technical and administrative control over the five Export Inspection Agencies (EIAs), one each at Channai, Delhi, Kochi, Kolkata and Mumbai established by the Ministry of Commerce, Government of India, under Section 7 of the Act for the purpose of implementing the various measures and policies formulated by the Export Inspection Council of India.

Export Inspection Council, either directly or through Export Inspection Agencies, its field organisation renders services in the areas of: Certification of quality of export commodities through installation of quality assurance systems (In-process Quality Control and Self Certification) in the exporting units as well as consignment wise inspection.Certification of quality of food items for export through installation of Food safety Management System in the food processing units. Issue of Certificates of origin to exporters under various preferential tariff schemes for export products.

4.9.1 Inspection and Certification

Three systems of inspection and certification have been recognized under the Fish and Fishery Products (Quality Control and Inspection) Order and Rules, 1988 notified by Government of India under Section 6 of the Export (Quality Control and Inspection) Act, 1963.

They are:

1. Quality Control and Inspection in Approved units (QCIA)
2. In process Quality Control (IPQC)
3. Self - Certification(SC)

Approval of the processing facility is mandatory for all the three types. The approval is accorded after inspection of the facility by a team of specialists and satisfying that it is having all the mandatory requirements.

A QCIA unit should have the prescribed programme support and processing controls. They should also keep the necessary records. However, the final product is subjected to inspection by the regulatory agency (EIA) to check whether it meets the laid down standards. Consignments meeting the standards are issued with certificate for export. Regular monitoring of the units is also done to ensure that the prescribed sanitary standards are maintained and hygiene requirements are met.

An IPQC unit should in addition, have a well-equipped microbiological laboratory and a qualified and approved technologist to carry out supervision of the processing and the quality control drills prescribed. This involves maintaining the facility in proper shape observing all the processing controls and inspection of the end product. The laboratory should be capable of conducting specialized tests for indole, heavy metals *etc.* besides the routine microbiological tests. Certificates for export are issued by the regulatory agency (EIA) based on records of the tests produced by the unit.

A SC unit besides meeting all the requirement of an IPQC unit should have arrangements for auditing the quality systems available in the unit. Such audits are designed to locate deficiencies in the system and ensure that time bound rectification's are carried out by the unit. Such units are authorized to issue certificates for their own products.

All the three types of units are kept under regular monitoring by the ELAs. QCIA units are visited every day by officers of the Agency while visits to IPQC units are undertaken once in a fortnight. In the case of self-certification units, visits by Agency officers are made on a further reduced scale.

The initial approval for QCIA, IPQC and SC units is for a period of three years.

Approvals are renewed after a systematic review of the quality systems in the unit.

Quality complaints on the product exported by the unit are investigated by the regulatory agency and if found genuine suitable actions are taken against the exporter/unit. This may involve withdrawal/suspension of the approval or even prosecution of the processer/exporter.

4.9.2 Fish Inspection Initatives and Resource Utilisation

The resources in the present inspection system by way of person, year and other infrastructure have not been fully made use of by the industry. The full utilization of inspection services can be made by way of training and other works related to Quality Assurance.

Domestic fish inspection services before they are supplied to the consumer can be taken up as a challenge. Both laboratory and inspection services can be made available for this. All work related to off shore inspection, which is not in existence, can be made available for this. All work related to off shore inspection, which is not in existence, can be taken up. This includes inspection and other sanitation programme of vessels, landing centres, primary and middle marketing centres.

Facility inspection, rating and certification on a regular interval can be made mandatory.

Introduction of ISO 9000 and HACCP in registered processing plants need increased resources and a certain amount of the present resources can be directed for this programme. ISO 900 or a system having equivalency, HACCP or a system having equivalency which are the requirements of EE countries and the US respectively in the future have to be given special emphasis and attention to sustain exports of seafoods.

A national programme for the implementation of the new policy in the inspection system by senior management is required. This involves formal and informal industries and other related departments. It also includes the efforts of study on job to maintain or upgrade technical competence. A national training programme strategy may be evolved for this activity.

Inspection of vessels and landing centres in the attitude can be directed based on a problem driven approach.

Interdepartmental activities will be required and these include job performed for or on behalf of other directorates, agencies or organizations other than inspection services. Exhibition, fair, preparation of books, manuals *etc.* are some of the other activities. The new approach should be directed to enhance and augment the testing facilities, essentially required for seafood inspection. This is also one of the criteria for the acceptance of the Indian system abroad. Facilities should be set up to reduce the time and cost on laboratory evaluation of seafoods without affecting the reliability and authenticity or the test results. The laboratory can be equipped with the state of the art analytical techniques reinforced with the most modern scientific instruments.

Collaboration with international organization such as FAO, Codex Alimentarius Commission *etc.* in planning and implementation of training programmes, not only with-in the country, but in the Asia Pacific region is another attempt to strengthen the quality assurance measures. Frequent exchange of resource personnel and technical information are being done to keep abreast of the changing scenario over the world.

Exchange of information in laboratory testing and change in methods are being done with US FDA lab, Canadian laboratories *etc.* to have the up to date information on lab procedures. All the above attempts will give the Indian inspection system a comfortable position among the international seafood inspection services with which we can be proud of having the best quality assurance measures, if properly implemented and monitored. With such increased transparency and communication between the authorities in importing countries and exporting countries, by the increased responsibility of industry in guaranteeing quality of the products and by providing option for the importing country to physically satisfy the system adopted in the selling countries the confidence of the consumer can be further increased.

4.10 Export Procedures

The following export documents are generally required for the export of fish and fishery products:

1. Proforma Invoice
2. Commercial Invoice
3. Packing List
4. Insurance Certificate
5. Mate's Receipt
6. Bill of Lading
7. Certificate of Origin
8. Bill of Shipping
9. Letter to Bank for negotiation/collection
10. Health Certificate

i. The Types of Export

1. FOB
2. C and F
3. CIF

ii. The Agencies involved in export are:

1. C and F agent
2. Shipping line
3. Surveyors
4. Customs
5. Port Trust
6. Chambers of Commerce
7. EIA/MPEDA
8. Exporter's Bank
9. Reserve Bank of India (RBI)
10. Insurance company

iii. The following are the procedures for the export of fish and fishery products.

1. Open a Current Account in any Commercial Bank in the Company's name.
2. Legally the company had to be registered under Industries Registration Act with the Registrar of Companies.
3. Apply through the Bank by depositing Rs. 1000/- by filling CNX form to get RBI code to deal foreign currency.
4. For registering the company's name
5. If the company is a single ownership, produce the Article of Memorandum and Register the company in any name.
6. If the company is a private limited, all shareholders will be called as Directors.

   Prepare the Article of Memorandum and give four names for registration and one name will be selected and the company will be registered legally under this name.
7. If it is a Public Limited, there will be Board of Directories. Prepare the Article of Memorandum and register the company's name.
8. Apply to Joint Chief Controller of Imports and Exports or Director General of Foreign Trade for getting Import and Export (IE) code by submitting the Registration Certificate and RBI Code.
9. Produce Land ownership Deed or Lease Agreement and lay-out of the plant to all Government authorities.
10. Apply to Pollution Control Board (PCB) to get clearance.
11. Apply for electricity and water.
12. Apply to Block Development Officers (BDO) for getting Revenue Department clearance
13. Prepare Project Report and apply to MPEDA for getting Export Certificate/ Registration Certificate by Submitting Project report along with land deed and lay-out of the plant.
14. Apply to get sales tax registration from both the Central and State Governments.
15. Start and complete the construction of the plant.
16. Apply to MPEDA to get certificate for processing hall, ice plant, peeling shed, laboratory and cold storage. MPEDA will issue temporary Registration/Export Certificate
17. Within a month after making one shipment, apply to MPEDA for permanent Registration/Export Certificate. Warehouse, if any, for storing the products has to be registered and certified by MPEDA.
18. Bacteriological analysis of processing plant water has to be done

19. Get certificate for weight and measures
20. Get Medical certificate for working labourers
21. Apply to EIA for their clearance and certificate and for getting packer's code number. For this, they require water sample analysis report, weights and measures certificate, medical certificate, and MPEDA certificate.
22. A panel consisting of Officials from EIA, MPEDA, CIFT and Local Chamber of Commerce or industries will visit the plant for inspection and approval of the plant.
23. Within three months of commencement of the plant, it has to be registered with the Inspector of Factories. They will regularize the labourers, safety measures and pollution control measures.
24. Apply for insurance for labourers and equipments.

4.11 Export Documentation

1. Purchase order from the buyer or buyer's representative
2. Buyer has to open letter of credit (LC) in buyer's bank through exporter's bank in exporter's name. This is called irrevocable sight LC.
3. Based on the buyer's specification, the products have to be packed.

i. Pre-shipment Document

1. Commercial invoice: Which contains material and packing specification, grade and value
2. Purchase order
3. Shipping bill-Prepared by the exporter or shipping agent. Submit shipping bill, commercial invoice and purchase order to customs for filling to do shipment. Customs will collect 1 per cent in the total value (½ per cent Agriculture tax and ½ per cent Fisheries tax)
4. Submit GR form (Exchange Control Form) to the exporter's bank to deal foreign exchange

ii. Post-shipment Document

1. Commercial Invoice
2. Bill of lading- receipt from shipping liners for evidence loading the material
3. Shipping bill
4. Packing list
5. Health Certificate
6. Certificate of Origin

7. Bacteriological report, if required

   Submit these documents to bank for negotiation

iii. International Code (INCO) Terms

1. Proforma Invoice: It may contain the following important details.
2. Ex-factory price (inclusive of normal profit)
3. Inland freight
4. Inland insurance charges
5. Forwarding agents charges
6. Port dues
7. Export duty
8. Pre-shipment inspection charges, if any
9. Freight charges
10. Bank charges
11. Marine insurance
12. Agency commission
13. Export Credit Guarantee Corporation (ECGC) premium, if any
14. Generalised system of preference (GSP) certificate charges, if any
15. Financing charges
16. Other charges, if any

It is always advantageous to send the proforma invoice to the foreign buyer, along with the prices, quantity including freight, terms of payment, mode of transport, insurance, delivery schedule, arbitration clause, *etc.* There are special invoice forms for many countries. The standard invoice forms may be available with chambers of commerce and other trade association.

1. **Mate Receipt:** As soon as the goods are loaded into the ship, the Master of the ship issues a "Mate Receipt"
2. **Bill of Landing:** On the basis of the "Mate Receipt" the bill of landing is issued by the Shipping Company. This some-times reflects the condition of the cargo shipped also. Hence, care should be taken that the cargo shipped without any damage at the time of shipment.
3. **Registered Exporter (REP):** In order to get the incentives like REP, announced by the Government, the exporter should get himself registered with the concerned Export Promotion Council or Commodity Boards.
4. **Export Proceeds:** When the documents are presented to the exporter's Bank, he gets the payment, of course subject to the terms of payment conditions.

5. **Document of Acceptance (DA):** This means that goods are supplied to the buyer on credit. It may range from 30 days to 120 days or sometimes more.
6. **Export Production:** As soon as the foreign buyer agrees to the quotation and places an order, production should be started and the foods should be got ready.
7. **Packaging:** It plays a vital role in export trade. Special labeling, or makes or the packages with identification marks, with special handling instructions may be required. Even in some cases, dimension of the package or material to be used are also mentioned in the sale contract.

iv. Standardised and Aligned Pre-shipment Export Documents

There are 16 Commercial Documents (of which 8 are "Auxiliary Documents") and 9 Regulatory Documents. Exporters are required to send the Principal Export Documents play a supportive role and are required for the preparation and procurement of the principal export trade. Auxiliary Documents play a supportive role and are required for the preparation and procurement of the principal export documents. Regulatory documents are associated with the pre-shipment stage of an export transaction and are issued by the Government authorities.

A. Principal Export Documents

1. Commercial invoice
2. Packing list
3. Bill of lading/Combined transport document
4. Certificate of Inspection/Quality Control (Where required)
5. Insurance certificate/Policy (In case of CIF export sales contract)
6. Certificate of Origin
7. Bill of Exchange
8. Shipment Advice

B. Auxiliary Documents

1. Proforma invoice
2. Intimation for inspection
3. Shipping instructions
4. Insurance declaration
5. Shipping order
6. Mate's receipt
7. Application for Certificate of origin
8. Letter to the Bank for Collection/Negotiation of documents

C. *Regulatory Documents*

1. Gate Pass-I Gate Pass-II Prescribed by Central Government
2. AR4/AR4A Form-Excise Authorities
3. Shipping Bill/Bill of export Custom Authorities
4. Export application/Dock challan/Port Trust Copy of shipping bill port trust
5. Receipt for payment of port charges
6. Vehicle ticket
7. Exchange control Declaration/GR/PP/Forms-Reserve Bank of India
8. Freight payment certificate
9. Insurance Premium Payment Certificate

4.12 Food Safety and Standards Authority of India (FSSAI)

Food Safety and Standards Authority of India (FSSAI) is an autonomous body established under the Ministry of Health and Family Welfare, Government of India. The FSSAI has been established under the Food Safety and Standards Act, 2006 which is a consolidating statute related to food safety and regulation in India. FSSAI is responsible for protecting and promoting public health through the regulation and supervision of food safety.

The FSSAI is headed by a non-executive Chairperson, appointed by the Central Government, either holding or has held the position of not below the rank of Secretary to the Government of India. The FSSAI has its headquarters at New Delhi. The authority also has 6 regional offices located in Delhi, Guwahati, Mumbai, Kolkata, Cochin and Chennai. 14 referral laboratories notified by FSSAI, 72 State/ UT laboratories located throughout India and 112 laboratories are NABL accredited private laboratories notified by FSSAI.

FSSAI notifies proposed changes in standards for Fish and Fish Products

The FSSAI has proposed amendments in standards to the regulation dealing with Fish and Fish products in the Food Safety and Standards (Food Products Standards and Food Additives) Regulations, 2011. The proposed standards have been dealt with briefly here.

i. Standards Proposed for Frozen Fish Products

- The freezing process will not be considered complete unless it has reached the temperature of -18°C at the thermal centre.
- The water used for cooking, cooling, glazing must be potable water (IS: 10500) or clean seawater with same microbiological standards as potable water.
- The products shall be kept deep frozen so as to maintain the quality during transportation, storage and distribution.

The Product

- ✰ Must have characteristic pleasant aroma and flavour and shall be devoid of any objectionable off – taste smell or odour.
- ✰ Artificial colouring matter and firming agents other than edible common salt and vinegar cannot be used
- ✰ Packaging of pickled fish requires special attention.

vii. Frozen Minced Fish Meat

Is prepared from clean and fresh fish which do not show any signs of degradation and spoilage. The fish shall be gutted; the tail, entrails, bones, tips, skin, head and other non- edible portion shall be removed and eviscerated and cleaned in potable water. Once prepared the material will be quickly frozen at a temperature not exceeding -30°C in polyethylene wrappers and packed in waxed cartons in the minimum possible time. The quick frozen material shall be stored in the cold storage at a temperature not less than -23 °C.

ix. Freeze Dried Shrimp

Means the product freeze dried prawns and must be prepared from clean, wholesome and fresh prawns.

x. Frozen Clam Meat

Its shall be of following two types:

- ✰ Raw Frozen Clam Meat (RFCM)
- ✰ Cooked Frozen Clam Meat (CFCM)

Freeze dried prawns and frozen clam meat must not show any visible sign of spoilage and meat shall be firm. The colour and odour will be that of freshly caught variety of the fish being processed. Water used in the processing shall be of potable quality and shall contain 5 mg/kg available chlorine.

xi. Common Standards for Fish Products

- ✰ Only regulation permitted food additives may be used
- ✰ Hygienic preparation and handling will be in accordance with the guidelines provided in part-II of Schedule 4 of the Food Safety and Standards (Licensing and Regulation of Food Businesses) Regulations, 2011 and any other such guidance provided from time to time.
- ✰ The products covered in this standard shall comply with the Food Safety and Standards (Contaminants, Toxins and Residues) Regulations, 2011.
- ✰ The products covered in this standard shall confirm to the microbiological requirements as given in Appendix B of these regulations.
- ✰ The products shall comply with the packaging and labelling requirements as laid down under the Food Safety and Standards (Packaging and Labelling), Regulations, 2011 and shall apply to the pre-packaged product.

4.13 International Standards

i. ISO 9000 Series of Quality Systems and Total Quality Management

Quality is defined as the "totality of features and characteristics of an entity that is able to satisfy stated and implied needs". This definition refers to concepts of fitness or use, worth for money, customer satisfaction and conformance to requirements and are not restricted to any age or culture.

After the IInd world war, there was a significant change in the scale and diversity of industry in general, and small self contained processing units changed into units having integrated operations where individuals do not have any control over the end products. Inspectors were appointed not connected with processing jobs to segregate the good product and the bad which came to be known as Quality Control System. This process of produce, inspect, accept, reject or rectify is the basis replaced by a better cost effective concept of producing it right first time – and every time and this is the quality assurance system. Standardisation of Quality took place in United States of America during this time. In Great Britain also such situations prevailed and all over the world. Standardisation method was developed in United Kingdom and was mainly based on BS 5750. As a result of this a Committee of the International Standards organization under the Chairmanship of Canada, worked out to produce International Quality Standards (FAO, 1994).

They considered many national inputs and in 1987 produced a standard which was largely based on BS 5750. This standard is ISO 9000 which embraces ISO 9001, 9002, 9003 and 9004. These standards were revised in 1994 and titled as ISO 9000 family of standards.

The concept of Quality underlying ISO 9000 is meeting requirements. A product or a service therefore has Quality when it satisfies the needs of the customers. ISO 9000 is a system for managing all activities in an organization that affect quality. It has been said that with ISO 9000 Quality standards, efficiency and productivity become system driven. A documented system must direct service, design, production, inspection, training *etc.*

It was developed in 1987 and was known as ISO 9000 series of standards and was revised in 1994. The revised standards are known as ISO 9000 family of standards and its Indian equivalent is IS 14000.

In ISO 9000 family there are 5 main standards. They are ISO 9000-Part I to ISO 9004-Part I.

Standard

Refers to reasonably complete and widely applied: product specification that has been agreed nationally and internationally, usually regulatory and inspected and certified.

ii. ISO 9000 Systems

ISO 9000–I – Quality Management and Quality Assurance Standards part I Guidelines for selection and use.

ISO 9001 Quality Systems:- Model for quality assurance in design, developments, production, installation and servicing.

ISO 9002 Quality Systems:- Model for quality assurance in production, installation and servicing.

ISO 9003 Quality Systems:- Model for quality assurance in final inspection and testing only.

ISO 9004 Part I: - Quality Management and quality system elements – Part I – Guidelines.

iii. Outstanding Features

1. Produced by people who are conversant with the problems and failures which occur in industries.
2. Although there are three parts to the system *i.e.*, 9001-9003 with minor changes, the systems can be made workable.
3. There is little dictatorial in the standards, in most cases it requires the company to establish its own procedures.
4. Even though traditionally confined to Engineering industry with minimal interpretations can be applied to food process industry.

iv. Advantages

1. Need not rectify work every now and then.
2. The work force is committed to the quality standards.
3. Due to proper auditing, management and motivation the standards open up and communication established at all levels.
4. Since preventive action is taken at all levels, than curing action frustration is eliminated.
5. Increased profitability is conductive to job security.

v. ISO 9000 Requires a Documented Quality System Comprising of

1. A policy document mentioning the Company's intention.
2. Operating procedures which are instructions on equipment operation, inspection *etc.*
3. Specific instruction on drawing, quality plan etc
4. A quality document communicates the quality policy and objectives of the company to its staff and customers and also the responsibilities of the authorities.

vi. Assessment

Assessment of manufacture and quality systems is an appraisal to establish whether the system meets quality system standards. The assessment process involves a documentation review and physical checking to find out if quality management system (a) exists (b) is correctly implemented (c) is effective.

To conduct assessment a number of foreign and Indian accredited certifying bodies are operating in India. The certificate is issued for a period of three years.

vii. Benefits of ISO 9000

- ISO 9000 has universal acceptance. It is a marketing tool.
- It streamlines all quality activities. System enhances productivity.
- Reduces cost
- Gives a competitive edge
- Ensures consistent quality
- Improves the morale of one and all
- ISO 9000 should not be seen as the ultimate system to quality but seen as first step in continuing improvement and customer satisfaction.

viii. Elements of Quality Management System

The elements are requirements in a quality management system. Twenty elements are available in ISO 9001 whereas 9002 has only 19 elements since design control is not required. ISO 9003 is the least comprehensive quality system standard. It requires a quality system that is composed of only 16 elements. It is intended to serve as a model when specified requirements are to be assured solely at final inspection and test for a commodity supplier ISO 9000 is suitable.

ix. Registration to ISO 9000 Standards

A company will follow the following steps for obtaining registration to ISO standards.

1. Commitment by top management
2. Formation of ISO steering group
3. Training in ISO
4. Development of Quality manual
5. Start-up internal audits (for corrective action and to eliminate root cause of problems)
6. Selection of registrar
7. Preassessment (Consultant to reveal problems)
8. Modified Q system implementation
9. Assessment by a registrar

Table 4.11: EEC Directive and Standards

Chemical	
TVB-N	– Not more than 30 mg/100g
Histamine	– Out of 9 samples, average not more than 100 ppm
	– 2 samples (100 X 2) = 200 ppm.
(No sample should exceed 200 ppm; applicable to Scombroidae and Clupeidae)	
Heavy metals	
Mercury	– 0.5 ppm
Cadmium	– 0.5 ppm
Tin	– 250 ppm
Parasites	– No tolerance
Microbiological criteria (Cooked, ready-to-eat shrimp and crab meat)	
Salmonella	– Not to be detected in 25 g
S. aureus	– m = 100/g; M = 1000
E.coli	– m =10; M = 100
Shrimp (APC) (without shell)	– m= 50,000; M = 5,00,000
Crab meat (APC)	– m = 1,00,000; M = 10,00,000

m: Acceptable count; M: Boundary between marginally acceptable and unacceptable.

Table 4.12: Standards for Frozen Fish and Fishery Products (Japan)

Quality Parameter	*Maximum Permitted Limit*
Viable bacteria	105 per g
Coliforms	Negative
V.cholerae	N.D.
Sulphur dioxide	100 ppm
Antibiotics	Nil
Dieldrin	0.1
DDT	No tolerance

Sampling

Samples to consist of a minimum of 5 units representative of the lot; analysis may be carried out on a composite of all sample units.

Criteria for Action

A lot will be considered rejected if the sample value exceeds the action level; fish or fish products exceeding these guidelines may be permitted for export if they do not violate regulations of the importing country.Contaminant level based on edible weight.

Table 4.13: Canadian Guidelines for Chemical Contaminants in Fish and Fish Products

Contaminant	*Product Type*	*Action Level*
Mercury	All fish products (edible weight except swordfish)	0.5 ppm
Arsenic	Fish protein	3.5 ppm
Lead	Fish protein	0.5 ppm
Fluoride	Fish protein	150 ppm
2, 3, 7, 8 TCDD(Dioxin)	All fish products	20 ppt
DDT and metabolites (DDD and DDE)	All fish products	5.0 ppm
PCB	All fish products	2.0 ppm
Pyperonyl butoxide	Dried cod	1. ppm
Other agricultural chemicals or their derivatives	All fish products	0.1 ppm

Filth is a menace and many countries are aware of the presence of flies and other insects, insect fragments, hair *etc.* in their product. To ensure safety of the consumers many countries made standards to restrict filth. USFDA stipulated regulatory requirements and standards for filth in seafood products (Table 4.14).

Table 4.14: USFDA Regulatory Requirements

Parameter	*Permitted Limit*
Salmonella/Arizona	ND in 375 g
Listeria (cooked only)	ND
Staphylococcus aureus	100/g
Sulphur dioxide	100 ppm
Mercury	0.5 ppm
PCBS	2 ppm
DDT and its derivatives	5 ppm
Filth in fresh or frozen raw shrimp	Nil

i. Flies and other insects (Whole or equivalent)

1. Filth insect – 2 in a sample
2. Incident insect – 3 in a sample

ii. Insect Fragments

1. Filth insect fragments 5 Fragments (excluding setae) in 2 of 6 slabs (clearly identified as parts of the filth insect)
2. Large body parts of filth insects (*i.e.* Thorax, abdomen) 1 in 2 of 6 slabs
3. Unidentified fragments 15 in a sample

iii. Hairs

1. Rat or mouse: 2 of any size in a sample
2. Striated but not of rat or muse: 3 of any size in a sample

4.15 Codex

a. The Codex Alimentarius Commission (Codex Standards)

This is an International body affiliated to FAO. This body aims to develop food standards to be used all over the world with a view to protect consumer's health and ensure fair trade practice. The member countries formulate standards based on this. These standards include provisions in respect of good hygiene practices, food additives, contaminants, labeling and presentation and methods of analysis and sampling. Codex standards are intended to guide, promote elaborate and establish definitions and requirements in foods and harmonise trade between countries. Two Committees namely one, the Commodity Committee deals with standards for food products including fish and fishery products, where as the 2nd Committee namely General Subject Committee deals with hygiene, labeling, additives, analysis and sampling and general principles on handling and processing. Codex standards for fish and fishery products are given in Table 4.15 (WHO, 1993).

Table 4.15: Codex Standards Concerning Fish and Fishery Products

Name of the Standard	*Number and Year*
Standard for Canned Pacific Salmon	Codex Stan. 3-1981
Standard for Canned Shrimp or Prawn	Codex Stan 37-1981
Standard for Canned Tuna and Bonito in Water or Oil	Codex Stan 70-1983
Standard for Canned Crab Meat	Codex Stan 90-1981
Standard for Canned Sardines and Sardine-type Products	Codex Stan 94-1981
Standard for Canned Mackerel and Jack Mackerel	Codex Stan 119-1981
Standard for Quick Frozen Gutted Pacific Salmon	Codex Stan 36-1981
Standard for Quick Frozen Fillets of Cod and Haddock	Codex Stan 80-1981
Standard for Quick Frozen Fillets of Ocean perch	Codex Stan 50-1981
Standard for Quick Frozen Fillets of Flat Fish	Codex Stan 91-1981
Standard for Quick Frozen Shrimp or Prawns	Codex Stan 92-1981
Standard for Quick Frozen Fillets of Hake	Codex Stan 93-1981
Standard for Quick Frozen Lobster	Codex Stan 95-1981

Chapter 5

Process Water Quality and Wastewater Treatment

Water and ice are used for a number of purposes in seafood handling and processing. In seafood handling water is mainly used for washing the raw material and the containers in which it is packed. In seafood processing water is used for washing raw materials, utensils, equipments, processing halls *etc.* It is used for glazing frozen seafood, preparation of brine and ice manufacture. Water is also used in boilers and in heating/cooling systems. Ice is used for chilling of seafood during handling and at different stages of processing. Quantity of water used by seafood processing plants varies with the product. During processing of seafood 0.5 to 1.0 gallons of water per water to 1 kg pound of fish processed may be used. As per Export Inspection Council (EIC) of India procedure for fixing the capacity of the seafood processing plant at least 10 litres of water is required to process one kg of raw material and 2 kg ice is required to process 1 kg raw material and 1.5 kg ice if the plant is fully air conditioned.

In order to obtain approval for processing of seafood from EIC of India the water used by the seafood processing plant should satisfy the requirements as per Indian Standard IS:4251, Quality Tolerances for Water for Processed Food Industry. The water used by the plant should meet the specifications as given in the Council Directive 80/778/EEC relating to the Quality of Water Intended for Human Consumption in order to export seafood to European Union, as article 2 of the directive states that the directive is also applicable to water used in a food production undertaking, for the manufacture, processing, preservation or marketing of products or substances intended for human consumption.

negative interference. Turbid water gives an aesthetically unpleasant opaqueness to water. The colloidal particles available in turbid water make the task of disinfection more difficult and expensive as the colloidal particles may behave as a protective shield for microbes. Treatment steps for water for the removal of turbidity upto 1000 NTU are coagulation, sedimentation, rapid filtration and post chlorination. A special type of filtration procedure known as microstraining is especially effective in the removal of the very small particles thereby reducing turbidity.

Odour

Water used for processed food industry should have no odour as per the standard. Odour is recognized as a quality factor affecting acceptability of drinking water and foods prepared with it. Most organic and some inorganic chemicals contribute odour or taste. These chemicals may originate from municipal and industrial water discharges, from natural sources such as decomposition of vegetable matter or from associated microbial activity and from disinfectants or their products. The presence of hydrogen sulphide imparts odour to water. It is not unusual for bore-wells to deliver water containing hydrogen sulphide which is readily dissipated, on exposure to the atmosphere or by aeration. Slow sand filtration may remove odour and taste. Under certain conditions chlorination may irradiate odour and taste. Ozonization may remove colour, odour and taste from water. Granular activated carbon beds can be used to remove most of the odour and taste producing compounds from water.

pH

The permissible range of pH is 6.5 - 9.2 as per the standard. Natural waters usually have pH values in the range of 4 to 9, and most are basic because of the presence of bicarbonates and carbonates of the alkali and alkaline earth materials. Presence of salt of weak bases and strong acids (*e.g.* $CaCl_2$) decreases the pH level. The potable water treatment processes such as chemical coagulation, disinfection, softening, and corrosion control, are pH dependent. Lime can be used to raise the pH of water. The mere addition of a small dose of lime may remove free carbon dioxide thereby increasing the pH of water. Removal of excess alkalinity in water is effected by passing carbon dioxide through the water.

Total Solids

The tolerance limit of total solids in water used for processed food industry is 1000 mg/l as per the standard. The term 'solids' refers to solid matter suspended or dissolved in water. Water with high dissolved solids generally are of inferior palatability and may induce an infavourable physiological reaction in the transient consumer. Highly mineralized waters also are unsuitable for many industrial applications. Eventhough the presence of dissolved and suspended matter in water is permitted upto 1000 mg/l, suspended matter should not produce turbidity above the tolerance limit (10 units) specified in the standard. Ion exchange is most commonly used to demineralize water thereby reducing total solid content.

Electrodialysis is used on a moderately large scale for the removal of dissolved solids. Distillation reduces dissolved solid content to the range less than one to 500 mg/l, depending on the type of distillation. Reverse osmosis is also used for the removal of solids.

Total Hardness

The maximum permissible concentration of total hardness in water is 600 mg/l (as $CaCO_3$) as per the standard. Total hardness is defined as the sum of the calcium and magnesium concentrations, both expressed as calcium carbonate, in milligrams per litre. As per IS:3025 total hardness is computed from concentrations of the different metallic cations (other than alkali metals) in the sample, but most often the cations taken into account are calcium, magnesium, iron, aluminium, zinc and manganese. Calcium hardness does not cause any public health problem. In fact, hard water is apparently beneficial to the human cardiovascular system. Hardness of water causes soap wastage during washing and scale formation in boilers and heating systems. If hardness is greater than 200 mg/l, lime softening can be employed for removal of hardness. Lime/soda ash treatment may also be used for removal of hardness. Ion exchange is most commonly used to soften water.

i. Sulphate

The maximum permissible level of sulphate in water is 200 mg/l as per standard. Sulphate is widely distributed in nature and is relatively abundant in hard waters. Waters containing large amounts of sulphates possess a faint bitter taste. Sulphates of sodium and magnesium in excessive amounts cause gastrointestinal irritation. Water containing an unusual quantity of sodium chloride or sulphate is an indication that the water may be derived from the stratum which contains these salts or there is infiltration of seawater. Since reverse osmosis membranes removes contaminants except dissolved gases, it can be used to remove sulphates from water. As electroialysis is applied for the removal of dissolved solids, it can also be used for the removal of sulphates.

ii. Fluoride

The maximum permissible level of fluoride is 1.5 mg/l as per the standard. Fluoride is the natural mineral contaminant of water. Fluoride is generally found in a few types igneous and sedimentary rocks. Fluoride is present in certain bore-well waters. A fluoride concentration of approximately 1.0 mg/l in drinking water effectively reduces dental caries without harmful effects on health. Excessive consumption of fluoride through drinking water cause a dental disease known as fluorosis or mottling. This causes discolourtaion of teeth. Excessive regular consumption of fluoride may give rise to bone fluorosis, *i.e.*, deformation of bone and other skeletal changes. If fluoride concentration is greater than 1 mg/l, alum flocculation, sedimentation, lime softening or ion exchange can be used for the removal. Reverse osmosis may also be used for the removal of fluoride.

iii. Chloride

The maximum permissible level of chloride is 250 mg/l as per the standard. Chloride is one of the major inorganic anions in water. Seawater is highly rich in chloride content. Hence ground water in coastal areas show significant high level of chloride. Chlorides are not harmful or toxic to humans even at concentrations as high as 2000 mg/l. But waters containing even 250 mg/l of chloride may have a detectable salty taste if the cation is sodium. The typical salty taste may be absent in waters containing as much as 1000 mg/l, when the predominant cations are calcium and magnesium. A high chloride content may harm metallic pipes and structures. Chlorides from drinking water can be removed by desalting methods such as reverse osmosis and electrodialysis.

iv. Cyanide

The maximum permissible level of cyanide is 0.01 mg/l as per the standard. Cyanide is very toxic, especially at low pH. Metallocyanide complexes formed by the reaction of cyanide ion with heavy metals are extremely toxic. Cyanide is an important industrial chemical and hence it is present in many industrial effluents. Drinking water sources can be contaminated by cyanide by discharge of industrial water to surface waters or leakage of industrial effluents into the drinking water sources. Cyanides can be removed by ozonization. Cyanides from wastewater can be removed by the following methods. 1. Complexing with ferrous ion and subsequent absorption employing ion exchange resin 2. Conversion to nontoxic biodegradable compounds. 3. Reaction with copper salts and oxygen and adsorption of the resultant copper cyanide complex on activated carbon.

v. Selenium

The maximum permissible level of selenium is 0.05 mg/l. Selenium is considered toxic to man. Selenium is also considered as an essential element. The beneficial and adverse effects depend on the amount injested, the form of selenium, the presence of other trace elements *etc.* The selenium concentration of most natural waters is low but in bore water in seleniferous soils in semi arid are as accumulates large concentration of selenium. Microbial degradation of seleniferous organic matter releases selenium compounds to water. Soluble selenium may be reached from coal ash and fly ash at electric power plants that burn seleniferous coal. Selenium is used in electronics, ceramic and shampoos. Selenium from drinking water can be removed by 1) Coagulation (ineffective for Se^{+6}), 2) lime softening (ineffective for Se^{+6}), 3) ion exchange, 4) reverse osmosis and 5) electrodialysis.

vi. Iron

The maximum permissible concentration of iron is 0.3 mg/l. Though there are no harmful effects for human from drinking water containing iron and manganese, they are unacceptable from an aesthetic point of view. Soluble iron and manageable are found in many ground water because of reducing conditions which farour the soluble +2 oxidation state of these metals. When the water are exposed to air, turbidity, colour and precipitate are developed due to the oxidation of the metals. Nuisance conditions can occur when the concentrations of iron and manganese are exceeded 0.3 mg/l and 0.1 mg/l respectively. In the absence of complex forming ions, ferric iron is not significantly soluble unless the pH is very low. Elevated iron levels in water can cause stains in plumbing, laundry and cooking utensils. Iron in water can impart objectionable tastes and odours to foods. The basic method for removing both iron and manganese depends upon oxidation to higher insoluble oxidation states. The oxidation is generally accomplished by aeration; chlorine and potassium permanganate are sometimes employed as oxidizing agents. These metals can also be removed by addition of sodium carbonate or lime.

vii. Magnesium

The tolerance limit of magnesium in water is 75 mg/l as per the standard. Magnesium salts which are found in waters are magnesium bicarbonate, magnesium sulphate, magnesium nitrate and magnesium chloride. Small quantities of chlorides of calcium and magnesium are found in many waters. Increased quantities are found in waters contaminated by seawater. Concentrations of magnesium greater than 125 mg/l can have a cathartic and diuretic effect. The softening methods employed for removal of hardness can also be employed for the removal of magnesium from drinking water.

viii. Manganese

The tolerance limit of manganese in water is 0.20 mg/l as per the standard. Elevated manganese levels can cause stains in plumbing, laundry and cooking utensils.

ix. Copper

The tolerance limit of copper for water for processed food industry is 1 mg/l. Copper in excess of 1 mg/l may impart some taste to water. Copper is considered an essential trace element, some compounds are toxic.

x. Lead

The tolerance limit of lead for water is 0.1 mg/l as per the standard. Lead has no beneficial or desirable nutritional effects and is a cumulative poison. Lead in a water supply may come from industrial, mine and smelter discharges or from the dissolution of plumbing and plumbing fixtures.

xi. Chromium

The tolerance limit of chromium for water is 0.05 mg/l as per the standard. Microgram amounts of chromium are essential for maintenance of normal glucose metabolism, but chromium (Cr^{+6}) is known to be toxic. Chromium is used in alloys, in electroplating and in pigments. The chromium guidelines for natural water are linked to the hardness or alkalinity of the water *i.e.* the softer the water, the lower the permitted level of chromium.

xii. Zinc

The tolerance limit of zinc is 15 mg/l as per the standard. Zinc in water produces undesirable aesthetic effects. Concentrations above 5 mg/l can cause a bitter astringent taste and an opalescence in alkaline waters. Zinc most commonly enters the domestic water supply from deterioration of galvanized iron and dezincification of brass. Zinc in water also may result from industrial waste pollution.

xiii. Arsenic

The tolerance limit of arsenic for water for processed food industry is 0.2 mg/l as per the standard. Arsenic is a cumulative poison. Arsenic occurs naturally in sulphide minerals such as pyrite. It is used in alloys with lead, in storage batteries and in ammunition. Arsenic compounds are widely used in pesticides and in wood preservations. Aqueous arsenic may result from mineral dissolution, industrial discharges or the application of pesticides. Conventional coagulations using alum or iron can be used for the removal of As^{+5}. Lime softening is also very effective for As^{+5}. Coagulation and lime softening are poor to moderately effective for the removal of As^{+3}. As^{+3} removal can be achieved by coagulation and lime softening, if As+3 is oxidized to As^{+5} prior to treatment. Ion exchange resins can remove both forms of arsenic. Reverse osmosis and electrodialysis should be effective for arsenic removal as these methods are very effective on the removal of most dissolved solids.

xiv. Nitrate

The tolerance limit of nitrate (as N) for water for processed food industry is 20 mg/l as per the standard. Nitrate generally occurs in trace quantities in surface water but may attain high levels in some ground water. In excessive amounts, it causes the illness 'methemoglobnemia' in infants. Nitrate, nitrite, ammonia, organic nitrogen and nitrogen gas are bio-chemically inter-convertible and are components of the nitrogen cycle. Nitrates may be derived from the oxidation of organic matter of animal origin. The more efficient the process of sewage purification, the less is the amount of nitrogen as free ammonia, and the greater the amount of nitrogen as nitrites and nitrates. Nitrates from drinking water can be removed by ion exchange, reverse osmosis and electrodialysis.

xv. Phenolic Substances

A limit of 0.001 mg/l (as C_6H_5OH) has been imposed on water for processed food industry as per IS:4251. Natural phenols may be present in water. Synthetic phenols are attributed to wastes from chemical industry. Presence of phenol in water may produce taste, particularly when the water is chlorinated. Granular activated carbon can be used for the removal of phenolic substance from drinking water. Superchlorination removes phenol present in the drinking water. Ozonization also removes phenol from drinking water.

xvi. Cadmium

The tolerance limit of cadmium for water for processed food industry is 0.01

mg/l as per IS:4251. Biologically cadmium is a nonessential, nonbeneficial element and a cumulative poison. Cadmium occurs in sulphide minerals. The metal is used in electroplating, batteries, paint pigments and in alloys with various other metals. The solubility of cadmium is controlled in natural waters by carbonate equilibria. Drinking water containing excessive cadmium led to the occurrence of *itai- itai* disease among Japanese.

xvii. Mercury

The tolerance limit of mercury for water for processed food industry is 0.001 mg/l as per IS:4251. Biologically mercury is a nonessential or nonbeneficial element and a cumulative poison. Mercury is used in amalgams, mirror coatings, vapour lamps, paints, pesticides, fungicides *etc.* The mercurous salts are less soluble than the mercuric and consequently are toxic. Alkyl compounds of mercury are the most toxic. There is no indication that concentration of mercury in drinking water have contributed in any significant way to methyl mercury intoxication of general population.

Heavy metals such as copper, lead, chromium, zinc, arsenic, cadmium and mercury can be removed from water by a lime treatment, electrodiposition, reverse osmosis, ion exchange, cementation or activated carbon adsorption.

5.1.4 Tolerances for Radioactivity

Everyone is exposed to some natural radiation that comes from both cosmic rays and terrestrial sources. There are large geographic variations in the amount of natural background radiation. A small amount of this unavoidable background radiation comes from drinking water that contains radionuclides. The radioactivity in water originates from both natural sources and human activities. The latter include operations concerned with the nuclear fuel cycle, medicinal uses of radionuclides, industrial uses of radioisotopes, worldwide fall out from atmospheric testing of nuclear devices and enhancement of the concentration of naturally occurring radionuclides. Three kinds of adverse health effects that radiation could produce are developmental and teratogenic effects, genetic effects and somatic (chiefly carcinogenic) effects. Gross alpha and gross beta measurements are useful for screening to determine whether further analysis for specific radionuclitides is merited.

i. Alpha Emitters

The tolerance limit of alpha emitters is 10^{-9} µc/ml (micro coulombs) as per IS:4251. Alpha particles are positively charged particles identical with helium nuclei which cannot penetrate the unbroken skin, but if an element liberating them is deposited in any organ of the human body it will give off these particles and may cause extensive damage.

ii. Beta Emitters

The tolerance limit of beta emitters is 10^{-8} µc/ml (micro coulombs) as per the standard. Beta particles are electrons negatively charged of variable penetrating power. Most of them will pass through about one centimeter of animal tissue.

5.1.5 Additional Tolerances for Specific Operations

The operations for which tolerances specified are 1) cooling, 2) washing, flushing and general purposes, and 3) processing.

1. Water for Cooling Purposes

The characteristics under the sub group are total hardness and slime-forming organisms

i. Total Hardness

For waters which are recirculated and used the tolerance limit of total hardness (as $CaCO_3$) is 30 mg/l as per IS:4251. In once through and run to waste systems carbonate hardness should be absent. Carbonate hardness is the hardness caused by the carbonates and bicarbonate of metals other than alkali metals. Non-carbonate hardness is the hardness caused by salts other than carbonates and bicarbonates of metals other than alkali metals. When total hardness is greater than total alkalinity (as $CaCO_3$), carbonate hardness is equal to total alkalinity. When total hardness is equal to or less than total alkalinity, carbonate hardness is equal to total hardness.

ii. Slime-Forming Organisms

As per the standard slime-forming organisms should be absent in water used for cooling purposes. If slime-forming organisms are present in cooling water, these organisms may form slimes which affects the efficiency of cooling.

2. Water for Washing, Flushing and General Purposes

There are three parameters *viz.* total hardness, iron and manganese under the sub group.

i. Total Hardness

As per the standard the tolerance limit is 30 mg/l especially if used for washing with soap or other alkaline detergents.

ii. Iron

The tolerance limit of iron is 0.1 mg/l as per the standard.

iii. Manganese

As per IS:4251 the tolerance limit for manganese is 0.1 mg/l for water for washing, flushing and general purpose.

3. Water for Processing Purposes

There are two parameters *viz.* iron and manganese under this sub group. The tolerance limit is the same (0.1 mg/l) for both the parameters.

5.1.6 Additional Tolerances for Individual Food Industries

Additional tolerances are specified for ten individual food industries. The individual industries are 1) bakery 2) canning 3) citrus fruit 4) confectionery 5) dairy 6) edible oil refining 7) gelatin manufacture 8) meat packing 9) starch and corn products manufacture and 10) sugar refining. Since seafood processing industry is not included in the individual food industries, these tolerances are not applicable to seafood processing industry except in the case of canning of seafood.

5.2 Quality Tolerances in the Council Directive Relating to the Quality of Water Intended for Human Consumption (80/778/ EEC)

The quality requirements of water specified in Council Directive 80/778/ EEC contains 62 parameters. Out of the 62 parameters maximum admissible concentration (MAC) is specified for 42 parameters and maximum admissible value is given for 2 parameters MAC for residual chlorine is not given in the directive. But the directive states that substances used in the preparation of water for human consumption do not, either directly or indirectly, constitute a public health hazard. Hence buyers from European Union specified MAC for residual chlorine as 2000 mg/l, as residual chlorine may constitute public health hazards. Guide levels are specified for 29 parameters. In addition to the 62 parameters there are four parameters which are applicable only to softened water intended for human consumption. The 62 parameters in the EEC directives are classified into five group : A. Organoleptic parameters, B. Physico-chemical parameters (in relation to water's natural structure), C. Parameters concerning substances undesirable in excessive amounts, D. Parameters concerning toxic substances and E. Microbiological parameters.

5.2.1 Organoleptic Parameters

The organoleptic parameters are Colour, Turbidity, Odour and Taste. The significances of these parameters were already discussed.

5.2.2 Physio-chemical Parameters (In relation to water's natural structure)

There are fifteen parameters in the group B. They are temperature, hydrogen ion concentration, conductivity, chlorides, sulphates, silica, calcium, magnesium, sodium, potassium, aluminium,total hardness, dry residues, dissolved oxygen and free carbon dioxide. Out of the 15 parameters the significances of six parameters *viz.*; pH, chlorides, sulphates, magnesium, total hardness and dry residues (total solids) were already discussed. The significances of the other parameters are given below.

i. Temperature

The maximum admissible temperature (25°C) specified for water in the Council Directive is based on the average ambient temperature in Europe and is not applicable to Indian Climatic conditions. Temperature readings are used in the calculation of various forms of alkalinity, in studies of saturation and stability with respect to calcium carbonate, in the calculation of salinity *etc.* Identification of source of water supply such as deep wells, often is possible by temperature measurement alone. Change in temperature affects the quality of water in a number of ways:

1. The suitability of water for total human body immersion is greatly affected by temperature. Comfortable temperature range for swimming purposes is from 20 to 30°C.
2. Temperature affects the self purification phenomenon in water. Increased temperatures accelerate the biodegradation of organic material in water resulting in increased utilization of dissolved oxygen. Moreover, oxygen becomes less soluble as water temperature increases. Thus increase in temperature may lead to total oxygen depletion and obnoxious septic conditions.
3. Indicator organisms are affected by temperature. Both total and faecal coliform bacteria are destroyed more rapidly in the environment with increasing temperatures.
4. Temperature affects water treatment processes. Lower temperatures reduce the effectiveness of coagulation with alum and subsequent rapid sand filtration. Decreased temperature also decreases the effectiveness of chlorination. Increased temperature may increase the odour of water because of the increased volatility of odour-causing compounds. Odour problems associated with plankton may also be aggravated.
5. Temperature changes in water bodies can alter the existing aquatic community.

ii. Conductivity

Electrical Conductivity (EC) of any sample of water depends upon the electrolytes dissolved in it. The EC is proportional to the amount of such substances dissolved in water provided the water is very dilute. Hence any change in the mineral character of a water will be indicated by a change in the EC. But EC does not give details of different ions present in the water. The ratio of total dissolved solids (TDS) to EC is usually the same, if the source of the water is the same. But the ratio is in the range 0.55 to 0.7 for any water. Hence if the ratio of any water is outside the range, the correctness of the determination of TDS and/or EC is doubtful and hence reanalysis for these parameters is required.

iii. Silica

Silicon is usually reported as silica (SiO_2) when rocks, sediments, soils and water are analysed. The common aqueous forms of silica are H_4SiO_4 and H_3SiO_4. Dissolved silicate in fresh waters is derived directly from weathered silicate minerals. High concentrations of silicate are generally found in tropical fresh waters.

In water treatment compounds of silicon are used. Activated silica is used as a coagulant. Siliceous zeolite used for softening of water is produced synthetically from sodium silicate and sodium aliminate or processed from natural minerals. In the presence of magnesium it can form scale deposits in boilers and in steam turbines. Silicosis resulting from human exposed to silica dust has been a common occupational disease.

iv. Calcium

Calcium causes scale formation. Being an important contributor to hardness in water it reduces the utility of water for domestic use. Calcium can be removed from water by employing any of the softening methods.

v. Sodium

Sodium occurs generally in lower concentration than calcium and magnesium in freshwaters and makes its way in water through weathering of rocks. In saline and brackish waters its concentration is remarkably high. Its salts are highly soluble in water. Sodium compounds are used in many application, including caustic soda, salt, fertilizers, and water treatment chemicals. The ratio of sodium to total cations is important in agriculture and human physiology. The general population is not adversely affected by sodium, but various restricted sodium intakes are recommended by physicians for a significant portion of the population. A limiting concentration of 2 to 3 mg/l is recommended in feed waters destined for high-pressure boilers. Sodium can be removed from water by the hydrogen-exchange process or by distillation. Sodium can be removed by deionizer.

vi. Potassium

Potassium occurs in ground waters as a result of mineral dissolution from decomposing plant material and from agricultural run off. Potassium compounds are used in glass, fertilizers, baking powders, soft drinks, explosives, electroplating and pigments. Potassium is an essential element in both plant and human nutrition. Unlike sodium it does not remain in solution, but is assimilated by plants and is incorporated into a number of clay-mineral structures. Potassium occurs in natural waters in far lesser concentrations than calcium, magnesium and sodium and hence a higher level of potassium in water indicates a change in its original structure. Potassium can be removed by using deionizer.

vii. Aluminium

Aluminium occurs in the earth's crust in combination with silicon and oxygen to form feldspar, micas and clay minerals. Aluminium and its alloys are used for heat exchangers, aircraft parts, building materials containers *etc.* Aluminium potassium sulphate (alum) is used in water treatment process to flocculate suspended particles, but it may leave a residue of aluminium in the finished product. Aluminium is nonessential for plants and animals. Water containing 0.2 mg/l of aluminium or more is not suitable for use by kidney dialysis patients. Aluminium is toxic to brain and is implicated in Alzeimer's disease. Aluminium can be removed from water by using chitosan.

viii. Dissolved Oxygen

Oxygen dissolved in water is an index of physical and biological process going on in water. There are two main sources of dissolved oxygen in water: diffusion from air and photosynthetic activity within water. Diffusion of oxygen from air to water depends upon solubility of oxygen, while solubility of oxygen is influenced by factors like temperature, water movements, salinity *etc.* Photosynthetic activity may cause exceptionally high concentration of oxygen, while respiration and salinity may lead to oxygen in water below saturation level. Moreover dissolved oxygen is utilized by organic matter in water during decomposition resulting in lower percentage saturation of dissolved oxygen. Insufficient dissolved oxygen in the water causes the anaerobic decomposition of any organic materials present producing noxious gases such as hydrogen sulphide. Dissolved oxygen content of water can be increased by means of aeration.

ix. Free Carbon Dioxide

Rainwater contains about 0.6 mg of carbon dioxide (CO_2) per litre. This is due to the dissolution of CO_2 from the atmospheric air. When precipitated water percolates through the soil, CO_2 is again dissolved from the air in the soil. Respiratory activity of aquatic organisms is an important source of CO_2, in surface waters. A part of the dissolved CO_2 combines with water to form carbonic acid (H_2CO_3) and free CO_2 includes CO_2 and H_2CO_3. Total CO_2 is the sum of free CO_2,carbonate(CO_3") and bicarbonate (HCO_3') present in water. Water containing free CO_2 reacts with limestone or chalk of soil or sediment producing calcium bicarbonate. Calcium bicarbonate thus formed is in equilibrium with free CO_2 in water. If the CO_2 is removed from the water, the equilibrium is affected and in order to attain equilibrium calcium bicarbonate is decomposed to form calcium carbonate and CO_2. These reactions affects free CO_2 content in water. Free CO_2 is frequently present in water at high levels due to decay of organic matter. Surface waters normally contain less than 10 mg free CO_2 per litre, while some ground waters may exceed that concentration. The free CO_2 content of a water may contribute significantly to corrosion. The free carbon dioxide content can be reduced by aeration.

5.2.3 Parameters Concerning Substances Undesirable in Excessive Amounts

There are 24 parameters in the group C. Out of the 24 parameters the significances of 8 parameters *viz.* nitrates, phenols, iron, manganese, copper, zinc, fluoride and residual chlorine were already discussed. The significance of the other parameters are given below:

i. Nitrite

Both in the oxidation of ammonia to nitrate and in the reduction of nitrate, nitrite is an intermediate oxidation state of nitrogen. Such oxidation and reduction may occur in natural waters and water distribution system. Nitrite is used as a corrosion inhibitor in some industrial process water. Nitrite is rarely found in drinking water at levels over 0.1 mg/L. Total oxidized nitrogen is the sum of nitrate and nitrite nitrogen.

ii. Ammonium (NH_4^+)

Most ammonia (NH_3) in water is present as NH_4^+ rather than as NH_3. Almost all natural waters contain ammonium salts, at least in traces. Rain and snow always contain a trace, the first fall containing most. All fertile soils and all decaying vegetable and animal matter contain free and saline ammonia. Sewage is rich in ammonium ion since urine of man and animals yields large quantities of ammonium carbonate. Live organisms present in water produce free and saline ammonia. Nitrates in water may be reduced to NH_3 by the reducing action of ferruginous sands or of considerable length of metal pipes. As pathogens may be present in sewage, contamination of drinking water by sewage is harmful. The source of free and saline ammonia in drinking water may be due to contamination by sewage. Hence studies may be concluded to ascertain the source, if free and saline ammonia content exceeds

0.06 mg/L. In the case of pure river waters free and saline ammonia content is rarely more than 0.04 mg/L and any excess is probably due to sewage, industrial effluent or drainage from manured land. Even the very small quantity which is sometimes regarded permissible may be derived directly from sewage. Hence interpretation of the significance of ammonium content in a sample of water is difficult. Ammonium ion content in water has no effect upon wholesomeness provided it is not associated with pathogens. During chlorination of water ammonium ions present in water react with chlorine forming chloramines. Chloramines are less active germicidal agents than chlorine. Hence traces of free and saline ammonia (ammonium ion) cause considerable retardation of sterilization. In ammonia-chlorine process of sterilization (treatment of water with chloramines) NH_3 may be added to water in amounts varying from 0.05 to 0.5 mg/L. Excessive ammonium ion in water may affect the efficiency of this method of water sterilization. Ammonium can be removed from water by passing through deionizer, oxidation with chlorine or by passing through a bed of activated carbon.

iii. Kjeldahl Nitrogen

Kjeldahl nitrogen includes organic nitrogen and ammonia and can be determined by kjeldahl method. The method is employed to determine nitrogen in the trinegative state. Kjeldahl nitrogen does not include nitrogen, in the form of azide, azine, azo, hydrazone, nitrate, nitrite *etc.* Kjeldahl nitrogen can be removed by oxidation with chlorine or treatment with calculated amount of alum.

iv. ($KMnO_4$) Oxidizability

When a water is mixed with potassium permanganate ($KMnO_4$) in acid solutions, a certain amount of oxygen is taken up from the reagent by the oxidisable matter in the sample. The absorption is very less in very pure waters but is high in waters polluted with animal or vegetable matter. The organic matter is slowly oxidized by the permanganate, but certain inorganic substances such as ferrous salts, nitrites and sulphides react with potassium permanganate instantly. An oxygen absorbed figure of 0.5 to 1 mg/L at 27°C after 4 h is satisfactory, but water is considered suspicious if the value exceeds 4 mg/L in the case of peaty waters and 2 mg/L in the case of other types of waters. Oxidizability can be removed by oxidization with chlorine or treatment with calculated amount of alum.

v. Total Organic Carbon (TOC)

TOC is a measure of all carbon atoms covalently bonded in organic molecules. During determination of TOC the organic moleculars are broken down and converted to CO_2 and the CO_2 is measured quantitatively. TOC is a more convenient and direct expression of total organic content than biochemical oxygen demand (BOD), assimilable organic carbon (AOC) or chemical oxygen demand (COD), TOC, unlike BOD or COD is independent of the oxidation state of the organic matter. TOCs in drinking water range from less than 100μg/L to more than 25,000 μg/L.

Organic compounds in water may react with disinfectants to produce potentially toxic and carcinogenic compounds.

vi. Hydrogen Sulphide (H_2S)

H_2S may be present in bore-well waters. The odour of H_2S disappears quickly when the water is allowed to stand exposed to the air. The threshold odour concentration of H_2S in clean water is between 0.025 to 0.25 μg/L. H_2S may also be produced during passage of water through mains and pipes owing to chemical or biological reduction of sulphates. Its common presence in wastewaters comes from the bacterial reduction of sulphate, decomposition of organic matter and industrial waste. H_2S can be removed from drinking water by aeration or chlorination. Forced draught aeration is more effective than natural draught aeration or cascading because of the scrubbing action in addition to the oxidation process. pH reduction with acid to a value between 4 and 5 is advisable in order to obtain the best results with forced draught aeration. When sulphide is present in high concentrations, the element sulphur may be precipitated and must be removed by filtration. Aeration will not remove sulphide completely from water. If it is desired to reduce sulphide to zero, chlorination is required. In order to reduce the chlorine dosage chlorination should be carried out as a secondary process after the bulk of the sulphide has been removed by the other processes.

$$4Cl_2 + 4H_2O + H_2S \rightarrow H_2SO_4 + 8\ HCl$$

vii. Substances Extractable in Chloroform

The amount of substances extractable in chloroform gives an idea of oils present in water.

viii. Dissolved or Emulsified Hydrocarbons after Extraction by Petroleum Ether (Mineral Oils)

Petroleum ether is used to extract "oil and grease" from water. 'Oil and grease" is composed primarily of fatty matter from animal and vegetables sources and hydrocarbons of petroleum origin (mineral oils). Some of the defects of "oil and grease" in the environment are the following. If the domestic water is contaminated by oil and grease, it may affect the taste and odour of the water. Oils or petrochemicals are toxic to aquatic organisms. Surface waters should be virtually free from floating non-petroleum oils of vegetable or animal origin and petroleum derived oils, as the floating sheens affect the oxygen transfer to the water. Bioaccumulation of petroleum products presents public health problems such as tainting of edible aquatic species, accumulation of carcinogenic polycyclic aromatic compounds in the tissues of marine organisms *etc.*

ix. Boron (B)

Boron may occur in natural waters, but significant quantities are received through discharge of industrial wastes. The most common form of boron in natural water is H_3BO_3. Seawater contains approximately 5 mg B/L. Drinking waters rarely contain more than 1 mg/L and generally less than 0.1 mg/L. These concentrations are considered innocuous for human consumption. If boron is ingested in a large concentration, it can effect the central nervous system. Prolonged ingestion can result in a chemical syndrome known as borism.

x. Surfactants

Surfactants enter waters and wastewaters mainly by discharge of aqueous wastes from household and industrial laundering and other cleaning operations.

Another sources is the effluent of industries dealing with detergent manufacturing.

Surfactants tend to collect at the air-water interface. The form produced by the detergent is very stable and interferes with the activity of aerobic bacteria necessary to carry out the oxidation of organic matter in wastewater. Since surfactants are used for cleaning operations, the presence of surfactants in water indicates that impurities which are removed by cleaning operations may also be present in the water. In the environmental waters the surfactant concentration generally is below 0.1 mg/L.

xi. Other Organochlorine Compounds not Covered by Parameter No. 55

Other organochlorine compounds include haloforms (trihalomethones) which are toxic.

Chlorination of water containing organic substances is the major factor in the formation of haloforms in treated drinking water. Haloforms can be removed by passing through powdered activated carbon (Hoehn *et al.*, 1978). Removal of organic substances before chlorination of water by ferric sulphate, followed by flocculation and sedimentation reduces the formation of tritralomethanes considerably during subsequent chlorination.

xii. Phosphorus (P)

As the P is found in rocks is generally insoluble in water, P content in natural waters is low. The use of phosphate based fertilizers and detergents increases the chances of contamination of drinking water. Phosphates are used extensively in the treatment of boiler waters. Sewage effluents contain phosphates, as the average daily phosphate content of urine varies from 1 to 7 g expressed as PO_4. Phosphates stimulate the growth of algae and other aquatic plants. It interferes with the coagulation process in water treatment and with lime-softening of water. Chemically, phosphate is most commonly removed by precipitation. Precipitants used are lime, aluminium sulphate, ferric chloride *etc.* Among the precipitants lime is the most commonly used. Phosphate is also removed from solution by adsorption on activated alumina.

xiii. Cobalt (Co)

Co is used in alloys, in fertilizers and in porcelain and glass. Co dust is flammable and is toxic by inhalation. Co is considered essential for algae and some bacteria, nonessential for higher plants, and an essential trace element for animals. Acute toxic effects in animals have been observed only at daily doses greater than several mg of Co per kg of body weight and chronic Co toxicity has been observed in children taking Co preparations to correct anemia at daily dose of 1 to 6 mg/kg body weight (Gillies, 1978). The average abundance of Co in streams is 0.2 g/L and in ground waters it is 1 to 10 µg/L. The concentration of Co found in any natural water or drinking water is too low to cause any adverse health effects.

xvi. Suspended Solids

The term 'solids' refers to solid matter suspended or dissolved in water. Solids include suspended solids, the portion of solids retained in a filter and dissolved solids, the portion that passes through the filter. Suspended and colloidal matter is responsible for turbidity in water. The presence of suspended and colloidal particles in water makes the task of disinfection more difficult as these particles may behave as a protective shield for microbes. Suspended solids can be removed by sedimentation, coagulation and flocculation or sand filtration.

xvii. Residual Chlorine

Excess of chlorine, both free and combined, remaining in the treated water can be removed by the addition of chemicals such as sulphur dioxide and sodium thiosulphate or by filtration through a bed of activated carbon.

xviii. Barium (Ba)

The solubility of Ba in natural and treated waters is controlled by the solubility of $BaSO_4$, as these waters usually contain sufficient quantities of sulphate.

The main use of Ba is in mud slurries used in drilling oil and exploration wells. It is used in pigments, rat poisons, pyrotechnics and in medicine. The average abundance of Ba in US drinking waters is 49 μg/L and in ground waters it is 0.05 to 1 mg/L. A limit of 1 mg/L is recommended for Ba in drinking water supplies because of the effect on the heart and blood vessels.

xix. Silver(Ag)

Trace amounts of Ag are found in some natural waters and in a few community water supplies. It has not been detected at levels exceeding the maximum admissible concentration of 50 μg/L. Ag is widely used in photography, silverware jewelry, mirrors and batteries. Silver iodide has been used in the seeding of clouds.

Silver oxide to a limited extent is used as a disinfectant of water. The average abundance of Ag in US drinking water is 0.23 μg/L and in ground water it is < 0.1 μg/L. Silver is nonessential for plants and animals. Ingestion of Ag or silver salts by humans results in deposition of Ag in skin, eyes and mucous membranes causing argyria, a permanent blue-gray discolouration of skin and eyes.

5.2.4 Parameters Concerning Toxic Substances

There are 13 parameters in the group D, parameters concerning toxic substances. Out of the 13 parameters the significance of seven parameters *viz.* arsenic, cadmium, cyanides, chromium, mercury, lead and selenium were discussed. The significances of the rest *viz.* beryllium, nickel, antimony, vanadium, pesticides and related products and polycyclic aromatic hydrocarbons are given below:

Beryllium (Be)

Be occurs in nature in deposits of beryls in granite rocks. The average abundance of Be in ground waters and in drinking waters is less than 0.1 μg/L. The oxide and hydroxide of Be are relatively insoluble in water; the sulphate and chloride are very soluble but they hydrolyse quickly to the insoluble hydroxide. Hence the solubility of Be in water is very low and it is approximately 0.1 μg/L at pH 6. Be is used in high strength alloys of Cu and Ni and windows in X-ray tubes. It is non essential for plants and animals. Be is not likely to occur at toxic levels in ambient natural waters. The major human toxic hazard potential of Be is via the inhalation of Be containing fumes and dusts that might emanate from processing and fabrication operations.

Nickel (Ni)

Ni seldom occurs in nature in the elemental form. It is obtained chiefly from pyrrhotite and garnierite. Ni salts are soluble and can occur as a leachate from Ni bearing ores. The average abundance of Ni in streams is 1 μg/L and in ground waters it is < 0.1 mg/L. Ni is used in alloys, magnets, protective coatings, catalysts and batteries. It is suspected to be an essential trace element for some plants and animals. Ni does not pose a toxicity problem because absorption from food or water is too low. Occupational exposure to nickel compounds through respiratory tract increases the risk of lung cancer and nasal cavity cancer.

Antimony (Sb)

The average abundance of Sb in streams is 1 μg/L, and in ground waters it is < 0.1 mg/L. It is used in alloys of lead, in batteries, bullets, solder, pyrotechniques and semiconductors. Soluble salts of Sb are toxic.

Vanadium (V)

V is found in minerals such as vanadinite and patronite. Vanadium complex are present in coal and petroleum deposits. The average abundance of V in streams is about 0.9 μg/L, and in ground waters it is generally <0.1 mg/L. V is used in steel alloys and as catalyst in the production of sulphuric acid and synthetic rubber. It is considered non essential for most higher plants and animals. V may play a beneficial role in the prevention of heart disease. Vanadium pentoxide dust causes gastrointestinal and respiratory disturbances.

Pesticides and Related Products

As per the council directive 'Pesticide and related products are insecticides (persistent organochlorine compounds, organophosphorus compounds and carbamates), herbicides, fungicides and polychlorinated biphenyls (PCBs) and polychlorinated triphenyls (PCTs). Pesticides are employed for many different purposes. Insecticides are used in the control of insects; herbicides used against plants and fungicides against fungi. The organochlorine pesticides commonly occur in waters that have been affected by agricultural discharges. Several of the pesticides are bioaccumulative, stable, toxic and carcinogenic. From a long term environmental standpoint organophosphorus insecticides are superior to organochlorine compounds because organophosphorus compounds readily undergo biodegradation and do not bioaccumulate. PCBs are found principally in water supplies contaminated by transformer oils in which PCBs were originally used as a heat-exchange medium. These compounds are toxic, bioaccumulative, and extremely stable. PCBs have been found through out the world in water. Exposure to PCBs is known to cause skin lesions in humans and to increase liver enzyme activity. PCTs have almost identical characteristics to high chlorinated PCBs. It is used as a substitute to PCBs, PCTs have been detected in water. Most of the 'Pesticides and related products' can be removed from water by passing through activated carbon, packed tower aeration or treatment with chitosan.

Polycyclic Aromatic Hydrocarbons (PAHs)

Polycyclic aromatic hydrocarbons (PAHs) often are by products of petroleum processing or combustion. Many of these compounds are highly carcinogenic at low levels. They are relatively insoluble in water. They are encountered abundantly in the environment.

5.2.5 Microbiological Parameters

There are six characteristics in the group E, microbiological parameters. Out of the six parameters total bacterial counts for water in closed containers is not applicable to seafood processing industry and the significances of coliforms and total bacterial count were discussed. The significances of the rest are the following:

Fecal Coliforms, Fecal Streptococci and *Clostridium perfringens*

In certain circumstances coliforms can multiply outside the intestines. Hence the presence of coliforms in water does not confirm the fecal pollution of the water.

The presence of fecal coliforms confirms the fecal contamination of the water. Fecal *streptococci* and *Clostridium perfringens* (sulphite-reducing Clostridia) are also indicator organisms. *Clostridium perfringens* form spores which survive for a much longer time than vegetative organisms. All the organisms except sulphite-reducing Clostridia can be controlled by proper chloranition of water. It is difficult to remove sulphite-reducing Clostridia by superchlorination as the orgnism forms spores.

5.2.6 Minimum Required Concentration for Softened Water Intended for Human Consumption

There are four parameters in the group. Out of the four the significances of three parameters *viz.* total hardness, hydrogen ion concentration and dissolved oxygen were already explained. Significances of the fourth parameter *viz.* alkalinity is given below Alkalinity of water is due to hydroxide, carbonate and/or bicarbonate ions. During softening of water bicarbonate ions may be removed completely and the water may become aggressive. Ilence a minimum admissible concentration (30 mg/L as HCO_3') is specified for alkalinity.

Conclusion

Significances of the parameters in the standards are discussed. If the characteristics exceed the limits, treatments are necessary and hence treatments are also discussed. The quality of water depends mainly on its source of origin. Water for seafood processing industry is obtained from surface water sources and/or ground water sources. In general, contamination is more in surface water than in ground water. But in certain cases impurities such as iron, manganese and fluoride are present in excessive amounts in ground water.

Table 5.1: Physical and Chemical Tolerances in IS:4251

Characteristics	*Tolerance*
Colour (Hazen units), max.	20
Turbidity (units), Max.	10
Odour	None
pH	6.5 – 9.2
Total solids, mg/l, Max.	1000
Total hardness (as $CaCO_3$), mg/l, Max.	600
Sulphate(as SO_4), mg/l, Max.	200
Fluoride (as F), mg/l, Max.	1.5
Chloride (as Cl), mg/l, Max.	250
Cyanide (as CN), mg/l, Max.	0.01
Selenium (as Se), mg/l, Max.	0.05
Iron (As Fe), mg/l, Max.	0.3
Magnesium (as Mg), mg/l, Max.	75.0
Manganese (as Mn), mg/l, Max.	0.2
Copper (as Cu), mg/l, Max.	1.0
Lead (as Pb), mg/l, Max.	0.1
Chromium (as Cr^{6+}), mg/l, Max.	0.05
Zinc (as Zn), mg/l, Max.	15.0
Arsenic (as As), mg/l, Max.	0.2
Nitrate (as N), mg/l, Max.	20
Phenolic substances (as C_6H_5-OH), mg/l, Max.	0.001
Cadmium (as Cd), mg/l, Max.	0.01
Mercury (as Hg), mg/l, Max.	0.001

Table 5.2: Quality Requirements of Water Specified in Council Directive

Sl.No.	*Parameters*	*Expression of the Results*	*Maximum Admissible Concentration (MAC)*
A.	**Organoleptic parameters**		
1.	Colour	mg/L Pt/Co scale	20
2.	Turbidity	mg/LSiO_2 Jackson units	10
3.	Odour	Dilution number	2 at 12°C 3 at 25°C
4.	Taste	Dilution number	2 at 12°C 3 at 25°C
B.	**Physico-chemical parameters**		
5.	Temperature	°C	25
6	Hydrogen ion concentration	pH unit	The water should not be aggressive Maximum admissible value 9.5

Sl.No.	*Parameters*	*Expression of the Results*	*Maximum Admissible Concentration (MAC)*
7.	Conductivity	μs cm-1 at 20°C	NS*
8.	Chlorides	Cl mg/L	NS*
9.	Sulphates	SO_4mg/L	250
10.	Silica	SiO_2,mg/L	NS*
11.	Calcium	Ca mg/L	NS*
12.	Magnesium	Mg mg/L	50
13.	Sodium	Na mg/L	150
14.	Potassium	K mg/L	12
15.	Aluminium	Al mg/L	0.2
16.	Total hardness		
17.	Dry residues	mg/L after drying at 180°	1500
18.	Dissolved oxygen	Per cent O_2 Saturation	Saturation value>75 per cent except for underground waters
19.	Free carbondioxide	CO_2 mg/L	The water should not be aggressive
C.	**Parameters concerning substances undesirable in excessive amounts**		
20.	Nitrates	NO_3 mg/L	50
21.	Nitrites	NO_2 mg/L	0.1
22.	Ammonium	NH_4 mg/L	0.5
23.	Kjeldahl nitrogen (excluding N in NO_2 and NO_3)	N mg/L	1
24.	($KMnO_4$)Oxidizability	O_2 mg/L	5
25.	Total organic carbon (TOC)	C mg/L	NS*
26.	Hydrogen sulphide organoleptically	S μg/L	Undetectable
27.	Substances extractable in chloroform	mg/L dry residue	NS*
28.	Dissolved or emulsified hydrocarbons (after extraction by petroleum ether); Mineral oils	μg/L	10
29.	Phenols (phenol index)	C_6H_5OH μg/L	0.5 Excluding natural phenols which do not react to chlorine
30.	Boron	B μg/L	NS*
31.	Surfactants (reacting with methylene blue)	μg/L (laury sulphate)	200
32.	Other organochlorine compounds not covered by parameter No.55	g/L	Haloform concentrations must be as low as possible
33.	Iron	Fe μg/L	200
34	Manganese	Mn μg/L	50
35	Copper	Cu μg/L	NS*
36	Zinc	Zn μg/L	NS*

Sl.No.	Parameters	Expression of the Results	Maximum Admissible Concentration (MAC)
37.	Phosphorus	P_2O_5 µg/L	5000
38.	Fluoride	F µg/L 8-12°C 25-30°C	 1500 700
39.	Cobalt	Co µg/L	NS*
40.	Suspended solids	–	NS*
41.	Residual chlorine	Cl µg/L	**
42.	Barium	Ba µg/L	NS*
43.	Silver	Ag µg/L	10
D.	**Parameters concerning toxic substances**		
44.	Arsenic	As µg/l	50
45.	Beryllium	Be µg/L	NS*
46.	Cadmium	Cd µg/L	5
47.	Cyanides	CN µg/L	50
48.	Chromium	Cr µg/L	50
49.	Mercury	Hu µg/L	1
50.	Nickel	Ni µg/L	50
51.	Lead	Pb µg/L	50 (in running water)
52.	Antimony	Sb µg/L	10
53.	Selenium	Se µg/L	10
54	Vanadium	Vµ g/L	NS*
55.	Pesticides and related products		0.1
	Substances considered separately		
	Total	µg/L	0.5
56.	Polycyclic aromatic hydrocarbons	µg/L	0.2

Sl.No.	Parameters	Results: Volume of the sample in ml	Maximum Admissible Concentration (MAC)	
			Membrane Filter Method	Multiple Tube Method (MPN)
E.	**Microbiological parameters**			
57.	Total coliforms	100	0	MPN<1
58.	Fecal coliforms	100	0	MPN<1
59.	Fecal streptococci	100	0	MPN<1
60.	Sulphite-reducing Clostridia	20		MPN≤1

Sl.No.	Parameters		Results: Size of Sample in ml.	Maximum Admissible Concentration (MAC)
61	Total bacteria counts for water supplied for	37°C	1	NS*
	human consumption	22°C	1	NS*
62	Total bacteria counts for	37°C	1	20
	water in closed containers***	22°C	1	100

Sl.No.	Parameters	Expression of the Results	Maximum Admissible Concentration (MAC)
F.	**Minimum required concentration for softened water intended for human consumption**		
1	Total hardness	mg/L	60
2	Hydrogen ion concentration	pH	The water should not be aggressive
3	Alkalinity	mg/LH CO_3	30, the water should not be aggressive
4	Dissolved oxygen		The water should not be aggressive

* NS - Not specified; ** Values should be measured within 12 hours of being put into closed containers; ** Necessary measures should be taken to ensure that substance used in the preparation of water for human consumption do not constitute a public health hazard.

5.3 Fish Plant Sanitation

i. Prerequisites to HACCP

Hygiene standards and procedures usually described as Good Hygienic Practices (GHP) or GoodManufacturing Practices (GMP), have been in place for many years and constituted an essential tool in traditional food control. These concepts are still essential in a modern food control system by providing the basic environmental and operating conditions for production of safe food and thus being a requisite or foundation for HACCP in an overall food safety management programme. What is new is the concept of formalising the prerequisite programme along side HACCP and the legal requirement in some countries (USA) of documented monitoring of certain sanitation areas.

In the Code of Federal Regulation (FDA, 2001) it is outlined what is covered by current GMPregulations. These include basically all procedures and practices necessary to produce safe foods.

ii. Good Manufacturing Practices (GMP)

Those procedures for a particular manufacturing operation which practitioners of, and experts in, that operation consider to be the best available using current knowledge. There is no clear definition of the term Good Hygienic Practices (GHP). However, "food hygiene"has been defined by Codex (CAC, 2001) as "all conditions and measures necessary to ensure the safety and suitability of food at all stages of the food chain."

iii. Good Hygienic Practices (GHP)

All practices regarding the conditions and measures necessary to ensure the safety and suitability of food at all stages of the food chain. The terms GMP and GHP therefore basically cover the same ground and for the purpose of thisbook, the term GHP will mainly be used.

iv. Various Definitions of GHP or Prerequisite

According to the Draft Revision of the Recommended International Code of Practice for Fish and Fishery Products (CAC, 2000), the following aspects should be included in the prerequisiteprogramme:

1. Requirements for fishing vessels – design and construction
2. Requirements for processing facility – design and construction
3. Design and construction of equipment and utensils
4. Hygiene control programme
5. Personal hygiene and health
6. Traceability and recall procedures
7. Training.

An example of the common prerequisite programme is given in an appendix to the publication by NACMCF (1998). Additional to the points listed by Codex it includes: Supplier control, specifications for all ingredients, chemical control and conditions for receiving, storage and shipping of raw materials and products. In the present publication some of these additional points,*i.e.* supplier control and specifications of ingredients, will be included in the HACCP plan and not in the prerequisite programme.

According to the US-FDA's seafood HACCP regulation (FDA, 1995), processors are required to have key sanitary conditions written into Sanitation Standard Operating Procedures (SSOPs). As outlined, SSOP are equivalent to GHP.

5.4 SSOP – Sanitation Standard Operating Procedures

The documented GMP for hygiene and sanitation required to meet the regulatory requirements for food control in the USA. They are also required to monitor these conditions and practices, correct unsanitary conditions andpractices in a timely manner and maintain sanitation control records. Thus the sanitation control procedures are an integrated part of the seafood HACCP regulations, but not of the HACCP programme.

The SSOP should address at least the following conditions and practices:

1. Safety of water and ice
2. Condition and cleanliness of food contact surfaces
3. Prevention of cross contamination from unsanitary objects to food

4. Maintenance of facilities for personal hygiene
5. Protection of food and food contact surfaces from adulteration
6. Proper labelling, storage and use of toxic compounds
7. Control of employee health conditions
8. Exclusion of pests.

The written SSOP plan should explain the sanitation concerns, controls, in-plant procedures and monitoring requirements. This will demonstrate commitment to buyers and inspectors and also ensure that everyone from management to production workers understands the basics of sanitation.

In the European Union (EU, the prerequisite requirements are included in both 'horizontal' legislation such as the Hygiene Directive (EC, 1993) and 'vertical' or commodity-specific legislation such as the Directive specifying the requirement for fish processing (EC, 1991). Many activities can be considered part of a prerequisite programme depending on the product and the actual processing conditions. For this reason, it is unlikely that two processing facilities have identical prerequisite programmes.Although definitions of prerequisites and/or SSOPs refer mostly to operational conditions, there are also basic requirements to the processing plant and the processing environment. Thus the SSOPs are specifying the quality of the water, maintenance of hygiene facilities *etc.*, but it is equally important that the plant has access to enough water and hygiene facilities (quantitative aspects).Below is a list of key points and activities that need to be addressed in any prerequisite programme:

The Processing Plant

1. Conditions of premises facilities: Water, ice, steam (quantitative conditions) water treatment system (chlorination plant, waste water treatment)sanitary facilities and installations
2. Equipment: Boxes, containers, and machinery.

Operational conditions and procedures (GHP): Safety of water and ice (qualitative conditions):

1. Cleanliness of food contact surfaces
2. Prevention of cross contamination from insanitary objects to food
3. Maintenance of facilities for personal hygiene
4. Protection of food from adulterants
5. Safe storage and use of toxic compounds
6. Control of employee health conditions
7. Pest control
8. Waste management
9. Transportation

10. Traceability and recall procedure
11. Training.

A proper and well designed prerequisite programme allows the HACCP team to focus and concentrate on the hazards directly applicable to the product and the processing procedures without undue considerations and repetition of protection from hazards from the surrounding environment. It is important to point out that the prerequisite programme certainly relates to safety and therefore is an essential part of the total quality assurance programme. Thus part of the prerequisite programme (*e.g.* sanitation controls) must lend itself to all aspects of a Critical Control Point (CCP) such as establishing critical limits, monitoring, corrective actions, record keeping and verification procedures. However, occasional deviation from a prerequisite programme requirement would not by itself be expected to create a food safety hazard of concern. Therefore deviations from compliance in a prerequisite programme usually do not result in reaction against the product.

This is in contrast to a CCP, where any deviation from the established critical limits always leads to reaction against the product.

The prerequisite programme is a good starting point for companies who have a long way to go to implement a HACCP system. Practical experience has shown that if the general issues related to the prerequisite programme are dealt with first, the HACCP study will be much more straight forward and the resulting HACCP plan easier to manage. All issues related to GMP,hygiene and the environment will be dealt with in the prerequisite programme and only truly'critical' control points, essential to safety of the product will be included in the HACCP plan.

5.5 Disinfectants

All food contact surfaces should be adequately and routinely cleaned and disinfected. Cleaning and disinfection belong to the most important operations in today's food industries. In the US, the term sanitation is sometimes used to describe the disinfection process. In some cases, sanitation may refer to the whole cleaning and disinfection process.

i. Food Contact Surfaces are

1. Those surfaces that contact human food and those surfaces from which drainage onto the food or onto surfaces that contact the food ordinarily occurs during the normal course of operations
2. Typical food contact surfaces include utensils, knives, tables, cutting boards, fish boxes, conveyor belts, ice makers, ice storage bins, gloves, aprons, *etc.*

The cleaning and disinfection process can be divided into clearly distinct operations. However,these are linked firmly together in that the final result will not be acceptable unless all processes are carried out correctly.

ii. Sanitation

Sanitation means adequately treating food contact surfaces by a process that is effective in destroying vegetative cells of microorganisms of public health significance, and substantially reducing numbers of other undesirable microorganisms, but without adversely affecting the product or its safety for the consumer.

iii. Cleaning

In the preparatory phase, the processing area is cleared of remaining products, spills, containers and other loose items. Machines, conveyors, *etc.* are dismantled so that all locations where microorganisms can accumulate become accessible for cleaning and disinfection. Electrical installations and other sensitive systems should be protected against water and the chemicals used.

iv. Cleaning Precedes Disinfection

An effective disinfection can only be obtained after an effective cleaning

The desirable plant disinfectant would be characterized by the following properties:

1. It has sufficient anti-microbial effect to kill the microorganisms present in the available time and should have a sufficiently low surface tension to ensure good penetration into pores and cracks
2. It rinses freely from the plant, leaving this clean and free from residues which could harmthe products
3. It does not lead to development of resistant strains or any surviving microorganisms
4. It does not cause corrosion or other deterioration of the plant. It is recommended that the suppliers of machines *etc.* be asked before chlorine or other aggressive disinfectant are used
5. It is not hazardous to the operator
6. It is compatible with the disinfection procedure being used, whether manual or mechanical
7. If solid, it should be easily soluble in water
8. Its concentration is easily checked
9. It is stable for extended storage periods
10. It complies with legal requirements concerning safety and health as well as biodegradability
11. It is reasonably economical in use.

It will often be necessary to combine disinfectants with additives in order to obtain the required properties. The following are among the most widely used disinfectants and shall be described briefly.

a. **Chlorine** is one of the most effective and widely used disinfectants. It is available in several forms, for instance sodium hypochlorite solutions, chloramines and other chlorine containing organic compounds. Gaseous chlorine and chlorine dioxide are also used. Chlorinated disinfectants at a concentration of 200 ppm free chlorine are very active and have a cleaning effect. The disinfectant effect is considerably decreased when organic residues are present. The compounds dissolved in water will produce hypochlorous acid, HOCl, which is the active disinfecting agent, acting by oxidation. In solution it is very unstable, particularly in acid solution where toxic chlorine gas will be liberated. Furthermore, solutions are more corrosive at low pH.

   Unfortunately, the germicidal activity is considerably better in acid than in alkaline solution, thus the working pH should be chosen as a compromise between efficiency and stability.Organic chlorinated disinfectants are generally more stable but require longer contact times. When used in the proper range of values (200 ppm free chlorine), chlorinateddisinfectants in solutions at ambient temperatures are non-corrosive to high quality stainless steel, but they are corrosive to other less resistant materials.

b. **Iodophors** contain iodine, bound to a carrier, usually a non-ionic compound, from which the iodine is released for sterilisation. Normally the pH is brought down to 2-4 by means of phosphoric acid. Iodine has its maximum effect at this pH range. Iodophors are active disinfectants with a broad antimicrobial spectrum like chlorine. They are inactivated by organic material. Concentrations corresponding to approximately 25 ppm free iodine will be effective. Commercial formulations are often acidic making them able to dissolve scales. They can be corrosive depending on the formulation and they should not be used above 45°C as free iodine may be liberated. If residues of product and caustic cleaning agents are left in dead ends and similar places, this may, in combination with iodophores, cause very unpleasant "phenolic" off flavours.

c. **Hydrogen peroxide** and **peracetic acid** are effective disinfectants acting by oxidation and with a broad antimicrobial spectrum. Diluted solutions may be used alone or in combination for disinfection of clean surfaces. They lose their activity more readily than other disinfectants in the presence of organic substances and they rapidly lose their activity with time. They should be used in concentration of 200-300 ppm.

d. **Quaternary ammonium compounds** are cationic surfactants. They are effective fungicides and bactericides but are often less effective against Gram negative bacteria. To avoid development of resistant strains of microorganisms, these compounds should only be used by alternating with the use of other types of disinfectants. Due to their low surface tension, they have good penetrating properties and for the same reason,they can be difficult to rinse off. If quaternary ammonium compounds come into contact with anion active detergents, they will precipitate and become inactivated. Mixing or successive use of these two types of chemicals must therefore be avoided. They can be used in concentrations of 200 ppm on food contact surfaces.

Types of Disinfectants

Disinfectant	*Forms/Description*	*Advantages*	*Disadvantages*
Chlorine	Hypochlorites Chlorine gas Organic chorine, *e.g.*, chloramines	– Kills most types of microorganisms – Less affected by hard water – Does not form films – Effective at low temperatures – Relatively inexpensive – Concentration easily determined by test strips	– May corrode metals and weaken rubber – Irritating to skin, eyes and throat – Unstable, dissipates quickly – Liquid chlorine loses strength in storage – pH sensitive
Iodophors	Iodine dissolved in surfactant and acid	– Kills most types of microorganisms – Less affected by organic matter than some – Less pH sensitive than chlorine – Concentration determined by test strips – Solution colour indicates active sanitiser	– May stain plastics and porous materials – Inactivated above 50°C – Reduced effectiveness at alkaline pH – More expensive than hypochlorites – May be unsuitable for CIP due to foaming
Quaternary Ammonium Compounds	Benzalkonium chloride and related compounds, sometimes called quats or QACs	– Non corrosive – Less affected by organic matter than some – Residual antimicrobial activity if not rinsed – Can be applied as foam for visual control – Effective against *Listeria monocytogenes* – Effective for odour control – Concentration determined by test strips	– Inactivated by most detergents – May be ineffective against certain organisms – May be inactivated by hard water – Effectiveness varies with formulation – Not as effective at low temp. as some – May be unsuitable for CIP due to foaming
Acid-Anionic	Combination of certain surfactants and acids	– Sanitize and acid rinse in one step – Very stable – Less affected by organic matter than some	– Effectiveness varies with microorganism – More expensive than some – pH sensitive (use below pH 3.0)

Disinfectant	Forms/Description	Advantages	Disadvantages
		– Can be applied at high temperature – Not affected by hard water	– Corrode some metals – May be unsuitable for CIP due to foaming
Peroxy Compounds	Acetic acid and hydrogen peroxide combined to form peroxyacetic acid	– Best against bacteria in biofilms – Kills most types of microorganisms – Relatively stable in use – Effective at low temperatures – Meets most discharge requirements – Low foaming; suitable for CIP	– More expensive than some – Inactivated by some metals/organics – May corrode some metals – Not as effective as some against yeasts and moulds
Carboxylic Acid	Fatty acids combined with other acids; sometimes called fatty acid sanitizers	– Kills most types of bacteria – Sanitize and acid rinse in one step – Low foaming, suitable for CIP – Stable in presence of organic matter – Less affected by hard water than some	– Inactivated by some detergents – pH sensitive (use below pH 3.5) – Less effective than chlorine at low temp. – May damage non-stainless steel materials – Less effective against yeasts and moulds than some
Chlorine Dioxide	A gas formed onsite and dissolved in solution or by acidification of chlorite and chlorate salts	– Kills most type of microorganisms – Stronger oxidiser (sanitizer) than chlorine – Less affected by organic matter than some – Less corrosive than chlorine – Less pH sensitive than some	– Unstable and cannot be stored – Potentially explosive and toxic – Relatively high initial equipment cost
Ozone	A gas formed onsite and dissolved in solution	– Kills most type of microorganisms – Stronger oxidiser (sanitizer) than chlorine and chlorine dioxide	– Unstable and cannot be stored – May corrode metals and weaken rubber – Potentially toxic – Inactivated by organic matter (similar to chlorine)
Hot Water/ Heated Solutions	Water at 77-88°C	– Kills most types of microorganisms – Penetrates irregular surfaces – Suitable for CIP – Relatively inexpensive	– May form films or scale on equipment – Burn hazard – Contact time sensitive

CIP: Cleaning In Place.

5.6 Detergents and Cleaning Schedule

Cleaning and Disinfecting in Seafood Processing Cleaning and Disinfection (referred to together as sanitation) forms one of the major day– to–day controls of the environmental routes of food product contamination. Cleaning refers to the process of removing dirt and food particles from an object or surface, both of which support the germs that cause food poisoning, food spoilage and also attracts pests. Cleaning reduces contamination and helps keep products safe to eat.

Protects and maintains equipment - clean equipment works more efficiently *e.g.* chillers and freezers work better if they are free from ice and dust

Provides a safe workplace-dirty equipment can be slippery and dangerous *e.g.* greasy or wet floors, utensils. Numerous and costly cases of food spoilage have been traced back to the insufficiencies of these procedures.

5.6.1 Terminology of Cleaning Agents used in Processing Detergents

A Detergent is a cleaning agent that helps to remove dirt and grease from porous surfaces (such as fabrics, clothes, non-treated wood) and/or non-porous surfaces (such as HmetalsH, HplasticsH, treated wood). All detergents are made principally of soaps or surfactants. Surfactants when mixed with water, help remove dirt from surfaces by helping the water spread over the surface that is to be cleaned (this is called wetting). Detergents allow the water to get in to the dirt. This is called penetration. Keeps the dirt removed from the surface of the water rather than having it settle on another surface (called emulsification and suspension).

Detergents Must

- Detergents must remove the dirt
- Detergents must not leave harmful residues after rinsing
- Detergents must not corrode or damage plant, surfaces and utensils
- Detergents must be safe to use
- Detergents use must not result in scaly deposits if used over and over
- Detergents should be good value for money
- Using Detergents

Use the Right Concentration

Use the right water temperature - the cleaning power of the detergent increases as the temperature of the water it mixes with is increased. Water which is too hot can bake on the dirt and make it difficult to remove. Have the detergent on the surface to be cleaned for the right amount of time, called contact time.

5.6.2 Rinse off the Detergent

Ask the supplier of your cleaning agents if they are food grade and always keep the accompanying material safety data sheet on file.

i. Detergents

A detergent is a substance which is (i) used to enhance the cleansing action of water (ii) an emulsifier, which penetrates and breaks up the oil film that binds dirt particles, (iii) capable of wetting surface(s) to allow it to penetrate the soil deposits and break the soil into fine particles (deflocculation) and to hold them in suspension so that they do not redeposit on the cleaned surface(emulsification) (iv) must have good sequestering power to keep calcium and magnesium salts in solution. There are two types of cleaning detergents: alkaline or acid that are often formulated with surfactants, chelating agents, and emulsifiers to enhance the effectiveness of the detergents. The most effective detergents in the dairy today are formulated with alkaline solutions that have chelators and surfactants, a wetting agent, which helps them to float off.

ii. Alkaline Detergents

The alkaline detergents are generally comprised of basic alkali, polyphosphates and wetting agents. None of the basic alkalis, higher phosphates or wetting agents can meet all the requirements of a good cleaner when used alone. *Basic Alkalis:* The basic alkalis, such as soda ash, caustic soda, trisodium phosphate and sodium metasilicate form the bulk of most of the common dairy cleaners. Two or more of them are used in combination to overcome the weaknesses of a single compound and to give certain desirable properties to the blended product.

For instance, Caustic soda is high in germicidal action and dissolving action on milk proteins, but it lacks deflocculating and emulsifying power as compared with other alkalis. In addition, caustic soda is objectionable in jobs requiring cleaning by hand because of hand burning as it is the most corrosive alkali. *Soda ash,* is the most common constituent of dairy detergents today, and is the most inexpensive form of alkali. It is a poor water softener and has only fair deflocculating and emulsifying action. It has the advantage of being a good buffer. It is obvious that soda-ash cannot be used in large proportions in cleaners to be used in extremely hard water. *Trisodium phosphate,* have become a very popular constituent in cleaners because of its ready solubility and high deflocculating and emulsifying powers. When compared with metasilicate or soda ash, trisodium phosphate is also relatively corrosive on tin unless metasilicate is present as a protective agent in the mixture. Concentrations are sometimes limited to 0.5–1.5 per cent to minimize phosphate levels in wastewater. *Sodium metasilicate,* has high active alkalinity and excellent deflocculating and emulsifying properties. It, like trisodium phosphate, is only a fair water softener. The calcium and magnesium silicates formed in hard water are flocculent and insoluble in solutions. Although it is the strongest alkali next to caustic soda, it is relatively non-corrosive and has the property of protecting metals against corrosion by other alkalis. Metasilicate is very effective in holding the soil in suspension during the washing operation so that complete cleaning is possible.

iii. Acid Detergents

The use of acid detergents is commonly restricted to the removal of milkstone, water scale (calcium and magnesium carbonates). Acid detergents are more effective against bacteria than are alkaline detergents. The two most common types of acid detergents used are:

i. Nitric Acid

Nitric acid has biocidal properties when used either as a pure acid or in more stable, less hazardous mixtures with phosphoric acid. In addition, nitric acid attacks proteins. Nitric acid offers the added benefit of forming a protective layer of chromium oxide on the surface of food processing equipments made up of stainless steel which contains chromium, thus preventing the leaching of iron ions into the milk.

Commercially available aqueous blends of 5-30 per cent nitric acid and 15-40 per cent phosphoric acid are commonly used for cleaning food and dairy equipment primarily to remove precipitated calcium and magnesium compounds (either deposited from the process stream or resulting from the use of hard water during production and cleaning). It should not be used in >1 per cent concentration for stainless steel surfaces.

ii. Phosphoric Acid

Phosphoric acid is used widely as the basis of acid cleaning materials and finds greatest application in the removal of milk-stone and similar deposits on surfaces such as protein deposits. Its performance is greatly enhanced by adding an acid-stable surfactant, which promotes penetration of surface deposits and also assists in the process of rinsing at the end of the cleaning process. It often is used at a concentration between 2 and 3 per cent w/v phosphoric acid for cleaning. Small quantities of complex organic acids are often added to enhance its effectiveness.

Disinfectants

A disinfectant is a chemical agent which is capable of destroying disease causing bacteria or pathogens, but not spores and not all viruses. The main difference between a sanitizer and a disinfectant is that at a specified use dilution, the disinfectant must have a higher kill capability for pathogenic bacteria compared to that of a sanitiser. A well prepared washing programme must include the following:

1. Mechanical removal of dirt using a brush or suitable cleaning tool.
2. Rinsing with tempered or cold water.
3. Washing with hot water (40-70°C) with detergent.
4. Rinsing with tempered/cold water.
5. Disinfection of the clean surface according to the manufacturer's instructions.

Sanitisers

Sanitisers have been widely used in the food industry to decrease populations of pathogenic and spoilage organisms in food production and processing facilities.

In general, to sanitise means to reduce the number of microorganisms to a safe level. The main difference between a sanitiser and a disinfectant is that at a specified use dilution, the disinfectant must have a higher kill capability for pathogenic bacteria compared to that of a sanitizer. Sanitisers are used following cleaning. Sanitisers reduce the number of bacteria, but they do not eliminate them completely. Different sanitisers kill bacteria in different ways. Most sanitisers belong to one of 4 Categories.

1. Chlorine sanitisers kill the microorganism by getting in to the microorganisms cells and altering its life.
2. Quaternary ammonia sanitisers (QUATS) break down the microorganisms wall and they die very quickly.
3. Iodophors- an antiseptic or disinfectant that combines iodine with another agent, such as a detergent
4. Acid sanitisers- Acid sanitizers have a broad spectrum germicidal activity and are very cost- effective to apply e.g Peracetic acid

In addition, Ozone has been used. Ozone is effective for the treatment of both biological as well as non-biological contamination. Ozone is a well-known and proven oxidizing agent that will destroy even the hardiest microbiological organisms as well as oxidising the majority of organic components. In addition to its universal application, ozone will leave no residue or chemical traces after the cleaning process is completed therefore making it a safe medium for the treatment and decontamination of cleaning in place

Applications Used in the Food Industry

Using Sanitisers Always check the label before use and do not mix sanitiser with other chemicals.

- ☆ Use the right concentration
- ☆ Use the right water temperature
- ☆ Leave it on for the correct time
- ☆ Selecting a Sanitiser must not leave any harmful residues.
- ☆ Be safe to use.
- ☆ Be able to be stored undiluted and still be effective in killing bacteria. Not result in strains of microorganisms which are resistant to sanitisers.

Sterilisation

Sterilisation is the process of removing all living cells, micro-organisms, pathogenic bacteria and spores from a product. This is usually done by subjecting the product to dry heat or pressure steaming. Sterilants are specialised chemicals, such as formaldehyde, which are capable of eliminating all forms of microbial life, including spores. The term sterilant conveys an absolute meaning; a substance can not be partially sterile.

Important Note

Never allow untrained staff to carry out cleaning for you and ensure that there is a material safety data sheet to accompany all cleaning chemicals used. Make a list of all equipment and surfaces to be cleaned.

- Ensure that staff carrying out the cleaning are provided with suitable protective clothing.
- Decide on the frequency that cleaning is carried out, eg daily, weekly, monthly *etc.*
- Discuss who is responsible for carrying out this task.
- Ensure that the cleaning chemicals used, are of food grade material.
- Ensure that cleaning record sheets are signed off when cleaning has been completed.

Review your cleaning schedule at regular intervals to ensure that it remains suitable for your operation. The efficacy of your cleaning regime can be verified by swabbing the areas that you have cleaned and sending the swabs to an accredited laboratory for analysis. Reviewing the laboratory reports as part of your overall HACCP Plan will indicate if your premises are being cleaned sufficiently.

5.7 Cleaning in Place

Cleaning in place or CIP, refers to all those mechanical and chemical systems that are necessary to prepare equipment for food processing, either after a processing run that has produced normal fouling or when switching a processing line from one recipe to another. Cleaning in place means that cleaning takes place without dismantling the system.

Importance

CIP is an important component in guaranteeing food safety in food processing plants. Successful cleaning between production runs avoids potential contamination and products that don't meet quality standards. Carrying out CIP correctly - from design to validation - ensures secure barriers between food flows and cleaning chemical flows. It is also important that CIP is carried out effectively and efficiently, and contributes to an overall low total cost of ownership (TCO). From the point of view of food processing, any cleaning time is downtime - the equipment is not productive. Cleaning must also be carried out safely, because very strong chemicals are involved that can be harmful to people and to equipment. Finally, it should be carried out with the least impact on the environment by using minimal amounts of water and detergents and by maximizing the re-use of resources.

The food processing industry – whether involving milk, cheese, yoghurt drinks or Béarnaise sauce – benefits immensely from advanced technology that can control processing and protect food quality, from raw materials coming in to packages going out. How to clean a plant that has been processing food depends on the type of food that has been produced, and under which conditions. Processing temperature and running time affect how the equipment will be soiled. Efficient CIP will depend on your knowledge of how mechanical, thermal and chemical processes work on different types of soiling.

Cleaning In Place (CIP) system of cleaning the interior surface of pipelines, vessels, filters, process equipment and associated things without dismantling. Industries that require high level of hygiene rely on CIP and they include dairy, beverage, brewing, pharmaceuticals, processed foods and cosmetics. The efficiency of cleaning and sanitation of milk contact surfaces are widely influenced by many factors like the character of contamination, micro topography of surfaces, straightness of passage ways, compatibility of surface agents, application methods, speed of application and related speed of penetration in to biofilm structure. Depending on the processing practice and load of soiling on the process equipment, the cleaning solutions may be used for single cycle or recycled and reused for multi use. In multiple use, cleaning solutions are drained after a few to several hundred cleaning cycles. In the dairy industry, reuse and multi use CIP systems operate by circulating chemicals and water without taking the equipment apart. The effectiveness of cleaning is preconditioned by factors like chemical agent, mechanical power, temperature, and time of the procedure.

5.8 Waste Management in Fish Processing Industries

This document provides an introduction and orientation to fish processors on the concepts of fisheries wastewater characterization and to the various types of treatment that can be applied. Similar to most processing industries, fish processing operations produce wastewater, which contains active organic contaminating organisms in soluble, colloidal and particulate form. Depending on the particular operation, the degree of contamination may be small (*e.g.*, washing operations), mild (*e.g.*, fish filleting), or heavy (*e.g.*, bloodwater drained from fish storage tanks).

The magnitude of the problem created by these wastewater streams cannot be generalized, and is strictly related to the individual case, since the impact depends, *inter* alia on the strength of the effluent, the rate of discharge, and the assimilatory capacity of the receiving waterbody.As in most wastewaters, the contaminants present in fisheries wastewaters are an undefined mixture of substances, mostly organic in nature. Since a detailed analysis for each component is neither useful nor practically possible, most of the analyses give an overall measure of the degree of contamination present. In the text dig follows an estimation of organic content is discussed since this is one of the main pollution parameters.

5.9 Physico-chemical Parameters

5.9.1 pH

The pH itself is not a contaminant but is important as a characterization parameter since it may reveal contamination or indicate the need for its correction before biological treatment of the wastewater. Effluents from fish processing plants are seldom acidic and are usually close to 7 or alkaline. This value is generally due to the decomposition of the proteinaceous matter and emission of ammonia compounds.

5.9.2 Solids Content

Solids may be present either in dissolved or suspended form. Suspended solids are of primary concern since they are objectionable on several grounds: those that settle can do so in the wastewater ducts, reducing their capacity; or if they settle in the receiving waterbody they may affect the bottom- dwelling flora and the food chain; if they float, the light dig enters from the surface is reduced, and those that remain suspended reduce the amount of light that enters the water thereby affecting wildlife. The solids which settle are usually measured with an Imhoff cone (Figure 5.1) in which a known amount of water (*e.g.*, 1 litre) is placed. The solids which settle are estimated at fixed times, usually after 10 minutes and after 2 hours. The admissible amounts that can be discharged depend on each regulation, but discharge of wastewaters is usually not permitted if they contain solids winch settle after 10 minutes.

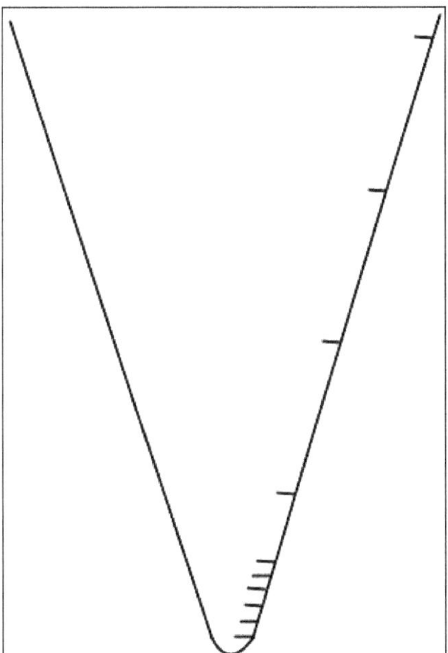

Figurc 5.1: Imhoff Cone Used to Measure Settling Solids.

The suspended solids are measured by passing a well-mixed sample through a fibreglass filter. The weight of suspended solids can be calculated by weight differences between the filter alone and the weight plus retained solids after drying. Soluble solids can be measured by gravimetry after evaporation of the filtrate of a sample of known volume, but are generally not checked even though in effluents with a low degree of contamination they can be significant. They depend not only on the degree of contamination but also on the quality of the supply water: in one analysis of fish filleting wastewater it was found that 65 per cent of the total solids present in the effluent were already in the supply water.

5.9.3 Temperature

With the exception of wastewaters from cooking and sterilization processes in a canning factory, fisheries wastewaters do not discharge wastes above ambient temperatures. The temperature of the receiving waterbody must not increase by more than 2°C or 3°C, since greater increases in temperature may affect the population balance and also reduce the solubility of oxygen, thereby threatening the survival of some forms of aquatic wildlife. Wastewaters from canning operations should be cooled if the receiving waterbody is not large enough to restrict the change in temperature to 3°C.

5.9.4 Odour

Odours in fisheries wastewaters are caused by the decomposition of the organic matter that emits volatile amines, diamines and sometimes ammonia. In wastewater that has become septic, the characteristic odour of hydrogen sulphide may also develop. Odours are very important in relation to the public perception and acceptance of any waste treatment plant. Although relatively harmless, they may affect people by inducing stress and nausea.

For measurement or estimation of odour, a test panel is exposed to odours diluted with clean air. The number of dilutions needed to reduce the odour intensity to its detectable limit is called as the detectable threshold odour concentration. The technique for determination of the threshold odour can be found in the Standard Methods, but as it is complicated and subject to errors due to adaptation of the test subjects, subjectivity and sample modification, it is seldom used.

5.10 Organic Content

The organic content of the wastewater can be estimated in several ways. The most common are the oxygen demand methods, although organic carbon measurement may also be used. The first estimate is the amount of oxygen that will be needed to stabilize the organic content of the effluent. The two most common methods are the biochemical oxygen demand and the chemical oxygen demand, which are discussed in detail in the following paragraphs.

5.10.1. Biochemical Oxygen Demand

The biochemical oxygen demand, also known as BOD, estimates the degree of contamination by measuring the oxygen required for the oxidation of organic matter by the aerobic metabolism of the microbial flora. In fisheries wastewaters, this oxygen demand originates mainly from two sources: the carbonaceous compounds which are used as substrate by the aerobic micro-organisms, and the nitrogen-containing compounds which are normally present in fisheries wastewaters, such as proteins, peptides and volatile amines. The most common procedures are described below.

i. Dilution Method

This is the most common procedure. It basically consists of diluting the wastewater (in a ratio that depends on its strength) with a nutrient (mineral salts) solution saturated with air, storing the dilution in airtight bottles, and measuring the dissolved oxygen at the start of the analysis and at periodic intervals. A five-day period is generally monitored and the BOD is then reported as BOD_5. For its analysis a simplified scheme is given below.

Simplified description of the BOD_5 test

- ✰ Obtain sample and transport immediately to laboratory
- ✰ Dilute samples with nutrient solution so that the maximum BOD is 6 mg/l
- ✰ Inoculate with adapted bacteria (see discussion in text)
- ✰ Fill the BOD bottles with the diluted effluent and seal them
- ✰ Fill blank bottles with dilution water only
- ✰ Determine by duplicate the dissolved oxygen (DO) in samples and in control at the start of test
- ✰ Incubate samples for 5 days in dark at 20°C
- ✰ Determine by duplicate the DO in samples and in control Calculate the net change in dissolved oxygen as: change in DO of sample, change in DO of control. The BOD_5 is the net change in DO multiplied by the dilution factor used in step 2.

Key points to be taken into account for a reliable BOD test will be highlighted. The first relates to the microbial population: since the BOD analysis involves biodegradation of organic matter by microbes, these must be present in the BOD bottles. In most fisheries effluents, the microbial count is such that this does not present a problem (a possible exception might be the wastewater from cooking or sterilization operations if the sample is taken either too hot or too close to the discharge from the autoclave). If the microbial count in the seed is not sufficient, the BOD may be under estimated because of the lag-time needed for the bacteria to proliferate. The same may happen if a non-adapted seed of bacteria is used. This under estimation may occur because of the extra time needed by the micro-organisms to adapt and/or grow to the extent needed to degrade the organic matter at a noticeable rate. This is the reason why the oxygen consumption in the BOD should be followed daily, and not only at the start of the analysis but for a

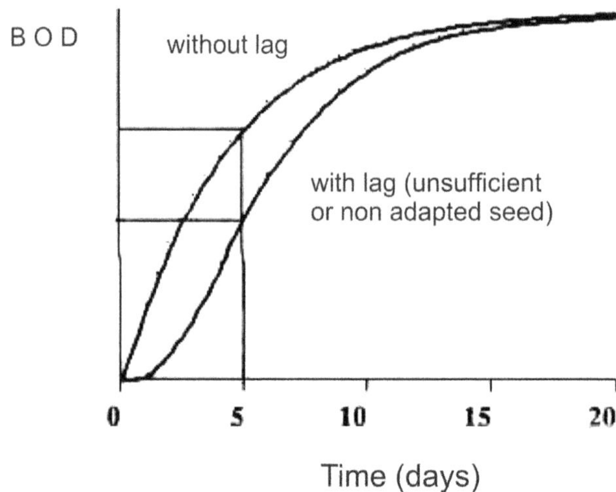

Figure 5.2: Typical BOD Curves.

5-day period. This second practice does not allow for detection of the pitfalls in the procedure mentioned above.

Another phenomenon that may occur is nitrification, since fisheries wastewaters contain appreciable amounts of proteinaceous compounds or products resulting from its degradation. This causes additional oxygen consumption. Although nitrification begins to take place after five days, it may occur earlier depending on the amounts and form of nitrogen present and the composition of the microbial flora.

To determine the BOD which is caused solely by carbonaceous matter, chemicals (such as allyl thiourea, methylene blue or 2-chloro-6-(trichloromethyl) pyridine) may be added to inhibit nitrification in the BOD analysis. However, since nitrification will impose an oxygen demand on the receiving waterbody, it should be taken into account as part of the total oxygen demand of a wastewater. The standard test calls for a 5-day incubation period at a constant temperature of 20°C in the dark. The latter is needed to avoid algae growth that might interfere if the bottles are exposed to light during incubation. As an alternative for highly contaminated wastes, lower incubation temperatures or shorter periods may be used, but the conditions of analysis should be clearly explained together with the results.

ii. Alternatives to the Dilution Method for BOD Determination

Although it is standard procedure, the dilution method involves inconveniences which have prompted the search for alternate, and simpler and more rapid procedures. Among the inconveniences are: the relatively long amount of time and the glassware needed for a reliable determination, the need to prepare several dilutions (with the consequent increase in the chance of errors) and the need for an acclimated seed.

An alternative to the dilution method is the respirometric method, which uses a bottle partially filled with the wastewater to be tested. The sample is continuously stiffed, and there is a reservoir with alkali (usually KOH). As the oxygen is consumed, CO_2 is liberated by the microbes and absorbed by the KOH and the pressure varies. The variations in pressure read by a manometer (Figure 5.3), allow for continuous reading of the oxygen demand. Another alternative is the use of an oxygen electrode within a bottle similar to that used in the dilution method. In this case, the bottle is filled with diluted sample, and the dissolved oxygen is measured continuously. Commercial versions of these alternatives are available.

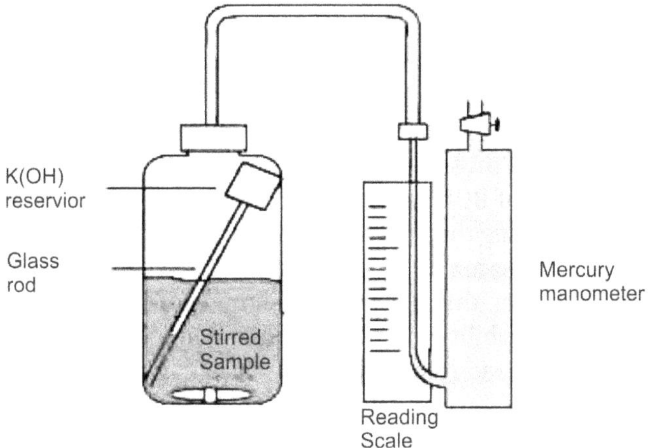

Figure 5.3: Manometric Device to Estimate the BOD of a Wastewater.

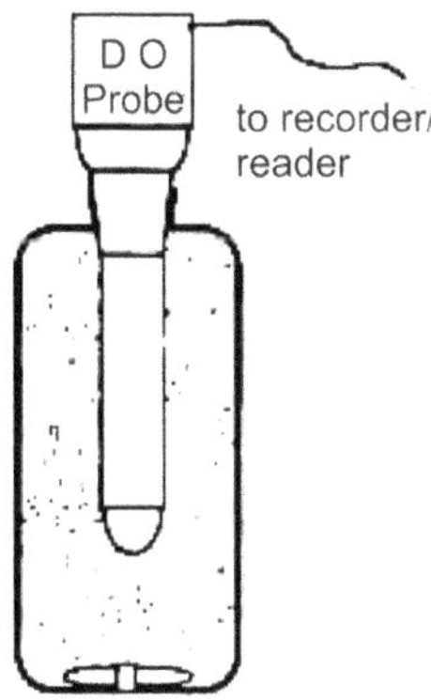

Figure 5.4: Dissolved Oxygen Electrode Assembled into a BOD Bottle.

One of the major inconveniences of the BOD_5 is the time it requires to obtain results. Five days render it unsuitable for control purposes of a treatment plant or for quick monitoring since the residence time of the wastewater in the treatment

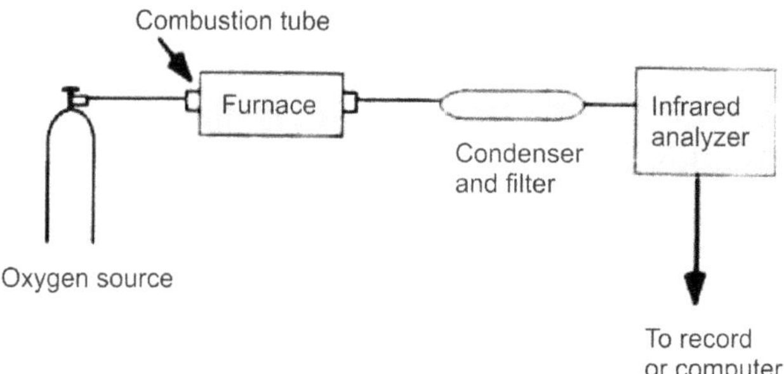

Figure 5.5: Diagram of a TOC Analyzer.

The signal recorded is proportional to the carbon dioxide formed which is directly proportional to the carbon concentration in the sample injected into the combustion tube. The oxidation of the sample proceeds either by heating to almost 1000°C or by intense UV radiation.

Since a wastewater contains carbonates or bicarbonates that are normally present in the supply water, these may be eliminated by acidification of the sample and purging with an inert gas prior to injection. One of the commercial versions available uses a different approach to differentiate between organic and inorganic carbon: it has two combustion tubes, one working at 150°C and the other at about 950°C. The first oxidizes the inorganic carbon, while the second oxidizes all the carbon present. The TOC is obtained from the difference between the readings. A major inconvenience for the widespread use of the TOC analysis is the relatively high cost of the apparatus.

Relationships between Estimates of the Organic Content

Since all of these procedures are based on oxidation of the substances present in the wastewater, in principle, a correlation may be established between their results. However, extreme care must be taken when developing these correlations since each of these analyses is affected by different factors. For example, the BOD_5 assay depends on many factors, such as adaptation of the seed and degree of dilution and proper pH; a minor fraction of the COD may be due to the presence of inorganic substances that reduce dichromate; there are compounds that are not degraded during the BOD_5 but are chemically oxidized such as toxic substances; there are substances resistant to oxidation by dichromate such as pyridine and acetic acid (although these are not normally present in fisheries wastewaters). Despite these factors, a correlation may be developed between two or more of the indicators of organic load. They are not applicable to situations different from that in which it was developed and, therefore, extrapolation is not reliable and should be periodically re-checked.

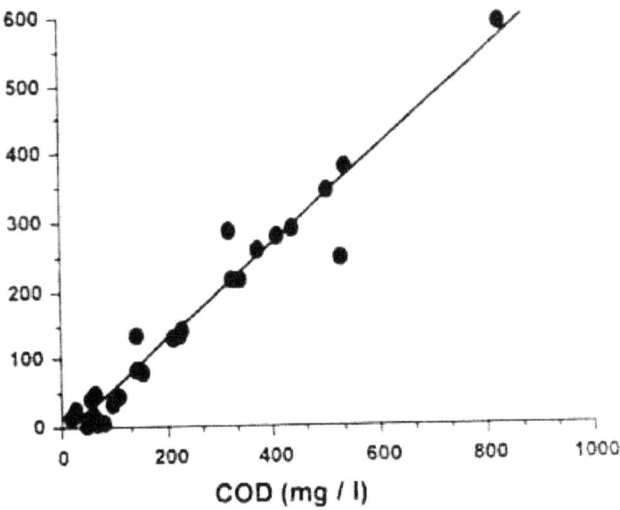

Figure 5.6: BOD_5 versus COD Relationship for Raw and Treated Fish Filleting Wastewater.

Oil and Grease

The presence of oil and grease in an effluent is mainly due to the processing operations (*e.g.*, canning) and, to a lesser extent, on the species being processed. In any case, they should be removed since they usually float and affect the oxygen transfer to the water and also objectionable from an aesthetic point of view. They may also cling to wastewater ducts and reduce their capacity in the long term. Generally, they are measured by extraction with solvent.

5.10.3 Nitrogen and Phosphorous

Both nitrogen (N) and phosphorus (P) are of environmental concern because they are nutrients, and if present in excess they may cause proliferation of algae (algal bloom) and affect the rest of the wildlife in a waterbody.Although N and P are normally present in the fisheries wastewater, their concentration is minimal in most cases. For biological treatment, a range of N to P in the order of 5 to 1 is recommended for proper growth of the biomass.

5.10.4 Characteristics of Fish Processing Wastewaters

The degree of pollution of a wastewater will depend on several parameters of which the most important are the operation being carried out and the fish species being processed. Considering only one type of operation, the operating routine in each factory also exerts strong influence on the wastewater characteristics. A particular parameter may present a wide range of numerical values. Table emphasizes that there is no substitute for a direct determination of contamination parameters in the effluent being considered. Moreover, given the variability that may be observed within one.

Table 5.3: Characteristics of Fish Processing Wastewaters

Effluent	*BOD*	*COD*	*Grease/ Oil*	*Total Solids*	*Suspended Solids*	*Ref. No.*
Finfish processing (manual)	3.32 kg/t		0.348 kg/t		1.42 kg/t	1
Finfish processing (mechanical)	11.9 kg/t		2.48 kg/t		8.92 kg/t	1
Patagonian hake filleting	327-1063 mg/l	550-1250 mg/l	8.3-79.9 mg/l			2
Herring filleting	3428-10000 mg/l		857-6000 mg/l			3.4
Tuna canning	6.8-20 kg/t		1.7-13 kg/t		3.8-17 kg/t	1
Sardine plant	9.22 kg/t		1.74 kg/t		5.41 kg/t	1
Blue crab plant	4.8-5.5 kg/t		0.21-0.3 kg/t		0.7-0.78 kg/t	1
Clam plant (mechanic)	5.14 kg/t		0.145 kg/t		10.2 kg/t	1
Clam plant (conventional)	18.7 kg/t		0.461 kg/t		6.35 kg/t	1
Fish meal plant	2.96 kg/t		0.56 kg/t		0.92 kg/t	1
Fish pumping water	2100-7400 mg/l		10-1504 mg/l	14.5-48.2 mg/l		5
Fish pumping water	3050-67200 mg/l		1300-17200 mg/l	18.4-64.9 mg/l		8
Blood water (fish meal plant)	23500-34000 mg/l	93000 mg/l	0%-1.92%	2.4%-6.3%		6.7
Stick water (fish meal plant)	13000-76000 mg/l		60-1560 mg/l	25-62 mg/l		5

5.10.5 Sampling

Particular attention should be paid to the representativity of samples. Since there is no general procedure applicable to all situations, the sampling programme must be designed for each situation. The location of sampling is usually made at or near the point of discharge to the receiving waterbody, but in the analyses prior to the design of a wastewater treatment, facility samples will be needed from each operation in the fish processing facility. With representativity in mind, samples should be taken more frequently when there is a large variation in flow rate, although wide variations may occur also at constant flow rate.

Commercial sampling equipment is available, although a simple continuous sampler may be constructed with regular tubing and glass carboys as shown in Figure 5.7 (Metcalf and Eddy Inc., 1979).

The particular procedure for sampling may vary depending on the parameter being monitored. Usually, the description of each standard technique includes

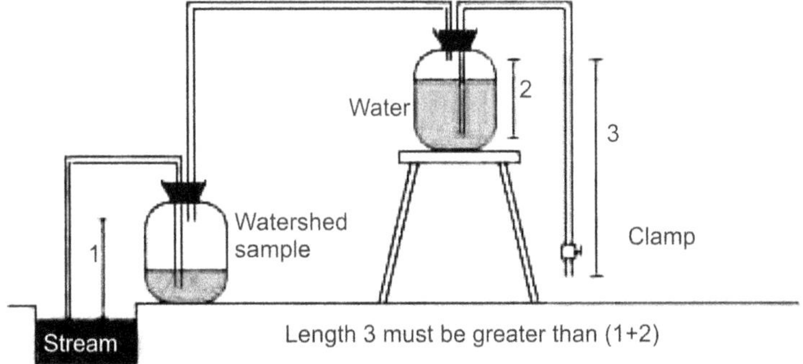

Figure 5.7: Continuous-Flow Sampling Device.

recommendation for sampling. If possible, samples should be analysed immediately since preservatives often interfere with the tests. In the case of fish processing wastewaters, there is no single method of sample preservation that yields satisfactory results for all cases, and all of them may be inadequate with effluents containing suspended matter. Care should be taken in blending the samples just prior to analysis since in almost all cases they will contain an amount of settleable solids. A case in which the use of preservatives is not recommended is that of the BOD_5. Storage at low temperature (4°C) may be used with caution for very short periods, and chilled samples should be warmed to 20°C before analysis. For COD determinations, the samples should be collected in clean glass bottles, and can be preserved by acidification to pH 2 with concentrated sulphuric acid.

For oil and grease, a separate sample should be collected in a wide-mouth glass bottle, well rinsed to remove any trace of detergent. The samples taken for organic nitrogen determination may be preserved similarly for COD. For the determination of solids, it should be checked that the suspended matter does not adhere to the walls, and the samples should be refrigerated at 4°C to prevent decomposition of biological solids. If phosphorus is to be analysed, preservation can be ensured by storing samples in well rinsed glass bottles (plastic bottles are not recommended) at -10°C and adding 40 mg/l of mercuric chloride.

5.10.6 Discharge Limits

In order for wastewaters to be discharged into the environment, there are a number of limitations for the concentrations of pollutants but there is no single criterion. In some cases, the limit is established by taking into account where the effluent is to be discharged. This type of criterion is also known as stream standards and is thought to be the more rational, taking into account the assimilative capacity and the intended use of the receiving waterbody. Another type of criterion is known as effluent standards. In this case, the maximum concentration of a pollutant (*e.g.*, in milligrammes/litre) or the maximum load (*e.g.*, kilogrammes/day) is discharged into a receiving waterbody.

5.11 Primary Treatment

Primary treatment is generally understood as the set of operations performed to remove floatable and settling solids. These solids are present in an effluent prior to secondary treatment in which biological and chemical processes are used to remove most of the remaining organic matter. In primary treatment only physical operations such as screening, sedimentation and flotation are used. The treatment used largely depends on the operations being carried out in the fish processing plant and on the requirements for disposal of the wastewater. Sometimes the only requisite is that no solids settle after 10 minutes, in which case simple screening and/or settling tanks with short residence times may be used. With more stringent norms, more elaborate processes such as flotation and biological treatment will be required.

5.11.1 Screening

By screening, relatively large solids (0.7 mm or larger) can be removed in a primary treatment facility. This is one of the treatments most commonly used by food processing plants as it quickly reduces the amount of solids being discharged. The simplest configuration is that of flow-through static screens, which have openings of about 1 mm. In some cases, such as streams with fish scales, they may require a scrapping mechanism to minimize clogging. The tangential screens are static but less prone to clogging due to their flow characteristics (Figure 5.8), since the wastewater flow tends to avoid clogging. Removal rates may vary from 40 to 75 per cent.

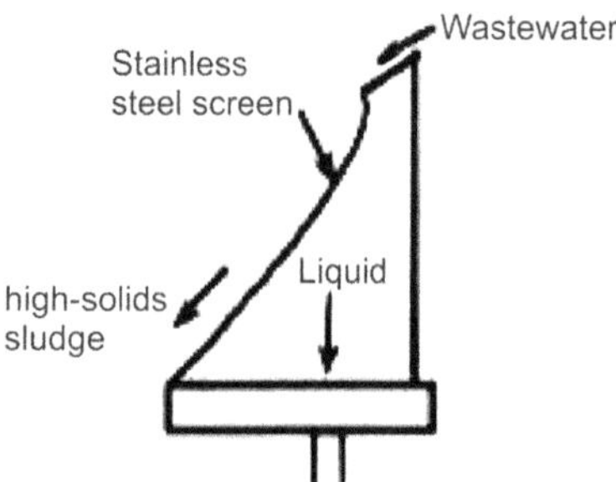

Figure 5.8: Diagram of an Inclined or Tangential Screen.

Rotary drum screens have also been used for fisheries wastewaters, although they are mechanically more complex. They consist of a drum which rotates along its axis, and the effluent enters through an opening at one end. Screened wastewater flows outside the drum. The retained solids are washed out from the screen into a collector in the upper part of the drum by a spray of the wastewater.

The screening media used in these devices is generally of stainless material, with openings varying from 0.7 to 1.5 mm. Since the fish solids dissolve in water as time proceeds, it is recommended that the waste streams be screened as soon as possible. By the same token, high intensity agitation of waste streams (such as pumping or flow-through valves) should be minimized before screening or even settling, since they cause breakdown of solids rendering them more difficult to separate.Both screening and settling are used in fish processing plants, screening being used more frequently in small-scale fish processing plants, together with very simple settling tanks. More elaborated sedimentation devices (discussed in the next section) are used in large-scale factories.

5.11.2 Sedimentation

Sedimentation is used to remove suspended solids present in the wastewaters. In fisheries wastewaters these include fish scales, portions of fish muscle and offal, the relative proportions varying with the particular process being used.

Sedimentation is based on the difference in density between the bulk of the liquid and the solid particles, which results in the settling of the solids present. The terms sedimentation and settling are often used interchangeably. This operation is conducted not only as part of the primary treatment, but also in the secondary treatment for separation of solids generated in the biological treatments such as activated sludge or trickling filters. Depending on the properties of solids present in the wastewater, sedimentation can proceed as:

- Discrete settling, if the wastewater is relatively diluted and the particles do not interact
- Flocculent settling, if the particles dig coalesce or flocculate are living particles of larger mass and faster settling rate. This is typical of untreated wastewater and is encountered in primary settling facilities
- Zone settling, it is also called hindered settling and occurs when the particles adhere together and settle as a blanket, forming a distinguishable interface with the liquid above it. This reaction occurs in secondary clarifiers for sludges of biological treatments.

Each case has different characteristics which will be outlined. For discrete settling, calculations can be made on the settling velocity of individual particles. In a settling tank, these move both downwards (settling) and towards the outlet zone with the waterflow (Figure 5.9). The particles that reach the bottom before the outlet will be separated from the effluent while the rest will pass through the settling tank. The critical velocity (vc) below in which a particle will be carried out of the tank is given by the depth of liquid (d), the volume of the tank (V) and the flow rate (Q):

$$v_c = d/(V/Q)$$

The ratio of V/Q is also known as the residence time of the liquid in the tank. It is called the overflow rate when vc is expressed in terms of volume of effluent per unit surface area of the tank per unit of time. This case may be present in fisheries wastewaters but is not thc most common.

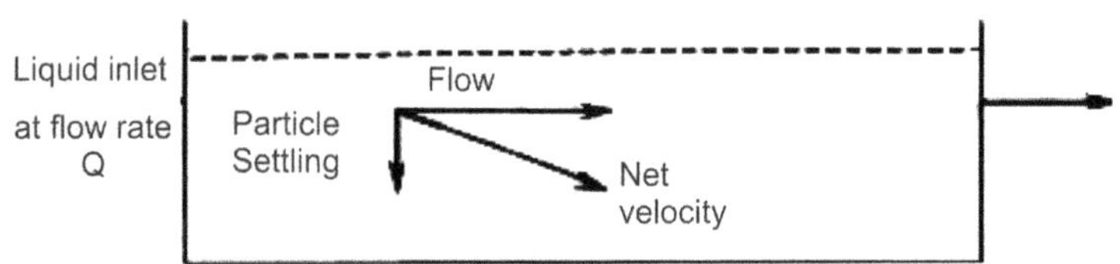

Figure 5.9: Schematics of Discrete Settling.

In the case of a flocculant suspension, the formation of larger particles due to coalescence depends on several factors, *e.g.*, the nature of the particles and the rate of coalescence. A theoretical analysis is not feasible due to the interaction of particles which depends among other factors on the overflow rate, the concentration of particles and the depth of the tank. A settling column is used to evaluate the. settling characteristics of a flocculant suspension (Figure 5.10). The same kind of column using only one sampling port can be used to study the discrete settling.

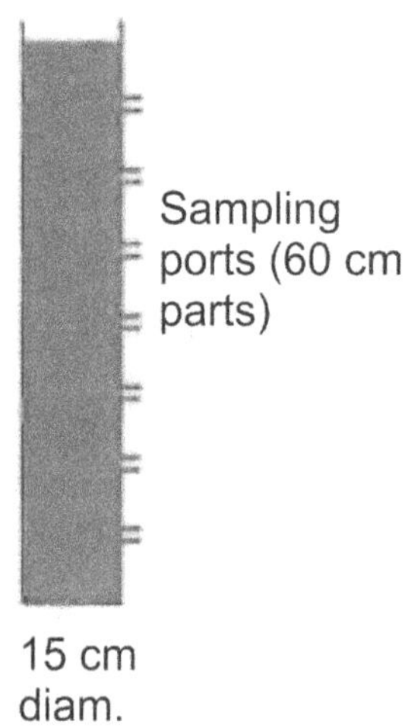

Figure 5.10: Laboratory Settling Column.

The zone (or hindered) settling, which occurs when the particles do not settle independently, is also studied by batch tests. In this case an effluent that is initially uniform in solids concentration (Figure 5.11), if allowed to settle, will do so in zones, the first of which is that of clarified water (1), below is the interfacial zone (2) in which the solids concentration is considered uniform. In the bottom a compact sludge develops in the so called compaction zone (4). Between (2) and (4), a transition zone (3) generally exists.

As time proceeds, the clarified effluent and compaction zones will increase in size while the two intermediates will eventually disappear. In some cases, further

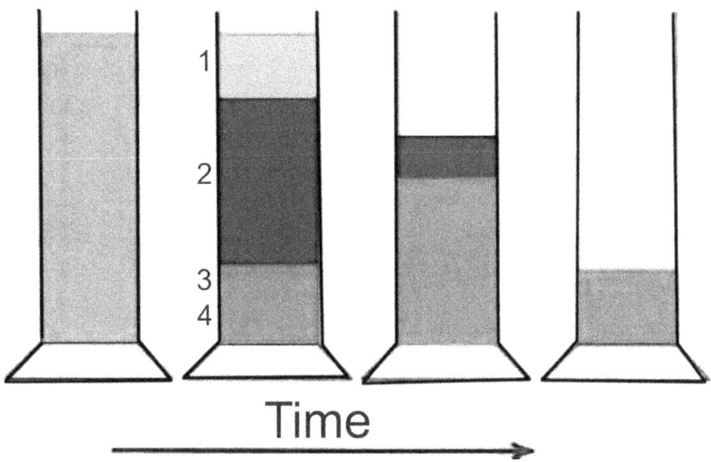

Figure 5.11: Diagram of a Zone Settling Process.

compaction may occur. The detailed design procedures for all these cases are beyond the scope of this document, and can be found elsewhere. The actual configuration of a sedimentation tank can be either rectangular or circular. Rectangular settling tanks (Figure 5.12) are generally used when several tanks are required and there is space constraint, since they occupy less space than several circular tanks.

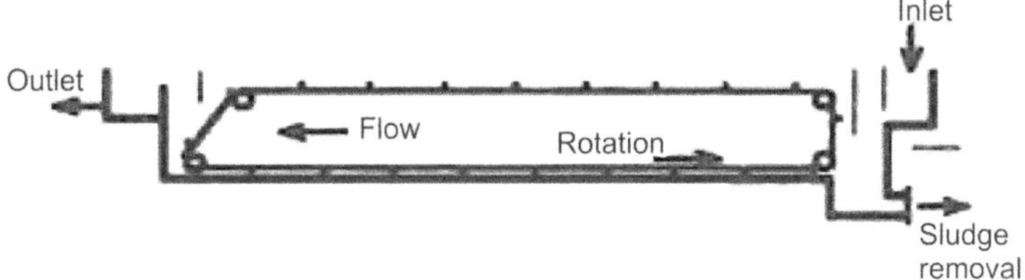

Figure 5.12: Diagram of a Rectangular Clarifier.

For removal of solids, a series of chain-driven scrapers are used: these span the width of the tank, are regularly spaced and move at 0.5 to 1 m/min. The sludge is collected in a hopper in the end of the tank, where it may be removed by screw conveyors or pumped out. Configurations exist in which the sludge is forced opposite to the flow, as shown here, but concurrent flow of the liquid and solids is also used.

The circular tanks are reported to be more effective. In these, the effluent circulates radially, the water being introduced at the periphery or from the centre. Figure 5.13 shows such a configuration. The solids removal are generally removed from near the centre, for which a slope of 10 per cent is required in the bottom of the tank. The sludge is forced to the outlet by two or four arms provided with scrapers which span the radius of the tank. For both types of flow, a means of distributing the

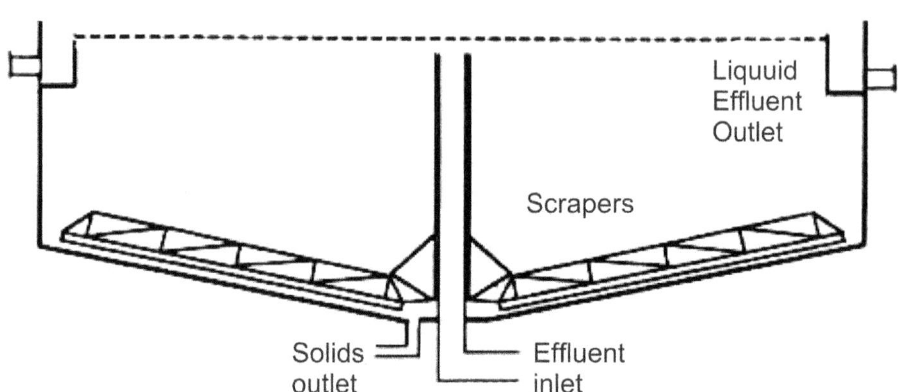

Figure 5.13: Diagram of a Rectangular Clarifter with Centre Feed.

flow in all directions is provided: for centrefed tanks there is a circular well, while for the rim-fed tanks a baffle is usually installed and the effluent enters tangentially. An even distribution of inlet and outlet flows is important to avoid short-circuiting in the tank that would reduce the separation efficiency.

A critical factor for selection of tank size is the so-called surface-loading rate, generally expressed as volume of wastewater per unit time and unit area of settler (m3/m2d). This loading rate depends on the characteristics of the effluent and the solids content, and can be determined from the settling tests described above. The retention time in the settlers is generally in the order of one to two hours, but the capacity of the tanks must be determined taking into account the peak flow rates so that good separation is also obtained in these cases. Some settling tanks, especially the larger ones, are provided with a mechanism for scum removal, since in biological wastes such as fisheries wastewaters, its formation is almost unavoidable.

In cases of small or elementary settling basins, the sludges can be removed using an arrangement of perforated piping placed in the bottom of the settling tank. The pipes must be placed regularly spaced (Figure 5.14), be of a diameter wide enough to be cleaned easily in case of clogging and the flow velocities should also be high enough to prevent sedimentation. These last two requisites are somewhat contradictory and a compromise is usually reached, using pipes of 5 cm in diameter, perforated with holes of 1-1.5 cm in diameter, 1 m apart. Flow in individual pipes may be regulated by valves. This configuration is best used after screening and is also found in biological treatment tanks for sludge removal.

An alternative to the above configurations for settling tanks is that of the inclined tube separators. These separators consist of tubes (although there are alternate designs that use plates close to each other) which are tilted.The concept is that, when a settling particle reaches the wall of the tube or the lower plate, it coalesces with another particle to give one of larger mass and higher settling rate.

The media are usually inclined 45°- 60°. They are also commonly used to upgrade existing settling tanks since they have a higher separation rate.

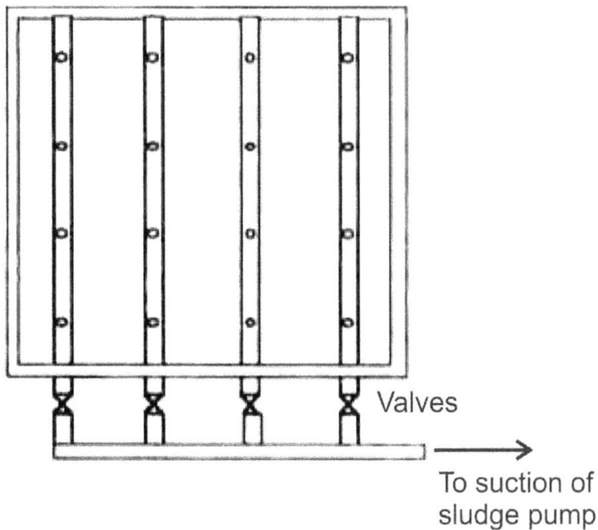

Figure 5.14: Pipe Arrangement for Sludge Removal from Settling Tanks.

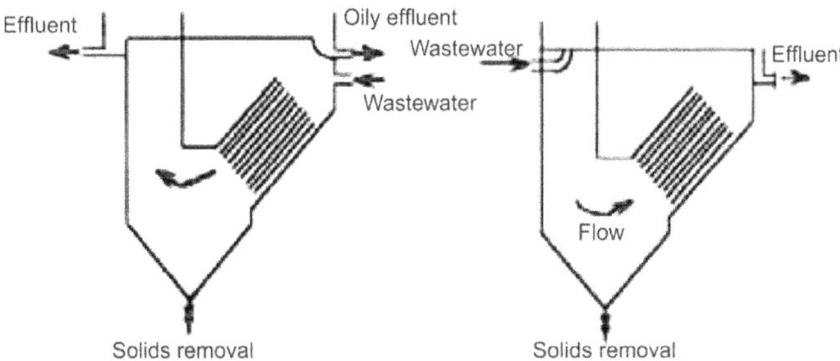

Figure 5.15: Typical Configurations for Inclined Media Separators.

5.11.3 Separation of Oil and Grease

Fisheries wastewaters contain variable amounts of oil and grease which depend on the process used, the species processed, and the operational procedure. To remove oil and grease, gravity separation may be used, provided the oil particles are large enough to float towards the surface and are not emulsified. If this is the case, the emulsion must be first broken, which in fish wastes may be achieved by adjustment of the pH. Heat may also be used but it may not be economical unless excess steam is available. Configurations of gravity separators of oil-water are similar to the inclined tubes or inclined plates separators discussed in the previous section. Variations of an original design of the American Petroleum Institute have been used in food processing wastes.

5.11.4 Flotation

Flotation is an operation that removes not only oil and grease but also suspended solids. It is discussed in this section since it is one of the most effective systems for suspensions which contain oil and grease. The most common procedure is that of dissolved air flotation (DAF), in which the waste stream is first pressurized with air in a closed tank. After passing through a pressure-reduction valve, the wastewater enters the flotation tank (Figure 5.16) where, due to the sudden reduction in pressure, minute air bubbles in the order of 50- 100 microns in diameter are formed. As the bubbles rise to the surface, the suspended solids and oil or grease particles adhere to them and are carried upwards. It is common practice to use chemicals to enhance flotation performance. As with coagulants (discussed later) these aids should preferably be innocuous, since these recovered solids are frequently used in animal feed formulations.

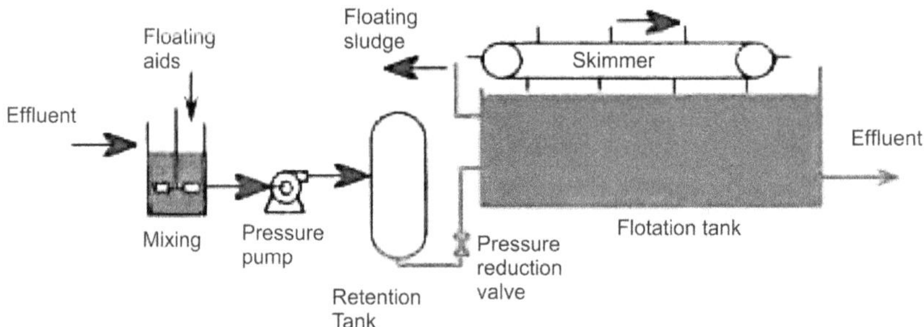

Figure 5.16: Diagram of a Dissolved Air Flotation System.

One alternate design involves the recycling of part (10-30 per cent) of the treated water (Figure 5.17). All systems contain a mechanism for removing the solids that may settle to the bottom of the flotation tanks, usually by a helical conveyor placed in the conical bottom. The main advantage claimed of DAF systems is the faster rate at which very small or light suspended solids can be removed in comparison with settling.

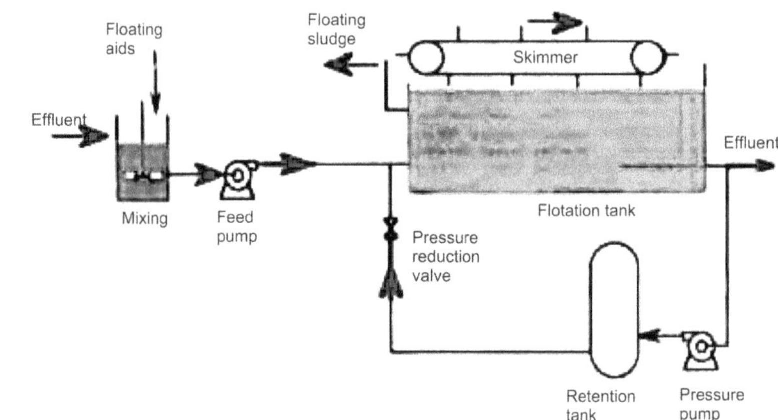

Figure 5.17: Diagram of a DAF System with Recycle.

Performance of DAF systems has been reported to be dependent on several factors, of which one of the most important is the solids concentration; higher solids content usually gives higher removal efficiencies. Other factors affecting the efficiency of the operation are the ratio of air to solids (A/S) defined as the amount of air released after pressure reduction and the amount of solids present in the wastewater. There is usually an optimum A/S which is determined by bench scale tests. Key factors in the successful operation of DAF units are the maintenance of proper pH (usually between 4.5 and 6, with 5 being most common to minimize protein solubility and break-up emulsions), proper flow rates and the continuous presence of trained operators.

In one case, oil removal was reported to be 90 per cent (Ilet, 1980). In tuna processing wastewaters, the DAF removed 80 per cent of oil and grease and 74.8 per cent of suspended solids in one case, and a second case showed removal efficiencies of 64.3 per cent for oil and grease and 48.2 per cent of suspended solids. The main difference between these last two effluents was the lower solids usually content of the second (Ertz *et al.*, 1977). Although considered very effective, DAF systems are probably not suitable for small-scale fish processing facilities due to the relatively high cost of a plant.

Another flotation system exists in which air is not dissolved but forced through the wastewater by surface aerators. This system generates air bubbles of larger size than DAF systems and no report exists about its application to fisheries wastewaters.

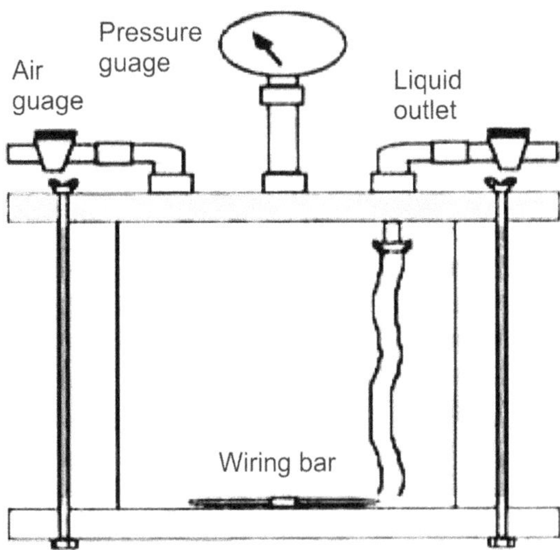

Figure 5.18: Laboratory Scale DAF Unit.

Prior to the design or selection of a DAF system, it is advisable to carry out laboratory experiments to evaluate its applicability and critical operating factors such as the air to solids ratio, the effectiveness of flocculants and the proper pH. This can be conveniently done in laboratory units such as that shown in (Figure 5.18). In these devices, water with or without chemicals and pH adjustment is introduced, and the pressure raised to the desired value. After mixing to saturate the liquid with air, pressure is released and the liquid flows to a graduated cylinder where time is allowed for separation. The detailed procedures for conducting the evaluations are available elsewhere.

5.12 Biological Treatment

The goal of all biological wastewater treatment systems is to remove the non-settling solids and the dissolved organic load from the effluents by using microbial populations. Biological treatments are generally part of secondary treatment systems. The microorganisms used are responsible for the degradation of the organic matter and the stabilization of organic wastes. With regard to the way in which they utilize oxygen, they can be classified into aerobic (require oxygen for their metabolism), anaerobic (grow in absence of oxygen) and facultative (can proliferate either in absence or presence of oxygen although using different metabolic processes). Most of the micro-organisms present in wastewater treatment systems use the organic content of the wastewater as an energy source to grow, and are thus classified as heterotrophes from a nutritional point of view. The population active in a biological wastewater treatment is mixed, complex and interrelated. By example, in a single aerobic system, members of the genera *Pseudomonas, Nocardia, Flavobacterium, Achromobacter* and *Zooglea* may be present, together with filamentous organisms (*Beggioata* and *Spaerotilus* among others). In a well-functioning system, protozoas and rotifers are usually present and are useful in consuming dispersed bacteria or non-settling particles. More extensive description and treatment of the microbiology of wastewater treatment systems are given elsewhere.

The organic load present is incorporated in part as biomass by the microbial populations, and almost all the rest is liberated as gas (carbon dioxide (CO_2) if the treatment is aerobic, or carbon dioxide plus methane (CH_4) if the process is anaerobic) and water. In fisheries wastewaters the non- biodegradable portion is very low.Unless the cell mass formed during the biological treatment is removed from the wastewater (*e.g.*, by sedimentation or other treatment described in the previous section), the treatment is largely incomplete, because the biomass itself will appear as organic load in the effluent and the only pollution reduction accomplished is that fraction liberated as gases.

The biological treatment processes used for wastewater treatment are broadly classified as aerobic (in which aerobic and facultative micro-organisms predominate) or anaerobic (which use anaerobic micro-organism. If the micro-organisms are suspended in the wastewater during biological operation, the operations are called "suspended growth processes", while the micro-organisms that are attached to a surface over which they grow are called "attached growth processes".

This section explains the principles and main characteristics of the most common processes in each case.

5.12.1 Aerobic Processes

In these, the reactions occurring can be summarized as:

Organic load + Oxygen + more cells + CO_2 + H_2O

In fisheries wastewaters, the need for addition of nutrients (the most common being nitrogen and phosphorus) seldom appears, but an adequate provision of oxygen is essential for successful operation of the systems. The most common aerobic processes are: activated sludge systems, lagoons, trickling filters and rotating disk contactors. These aerobic processes are described, together with the devices used for aeration.

i. Activated Sludge Systems

These systems originated in England in the early 1900's and earned their name because a sludge (mass of microbes) is produced which aerobically degrades and stabilizes the organic load of a wastewater. Figure 5.19 shows the layout of a typical activated sludge system.

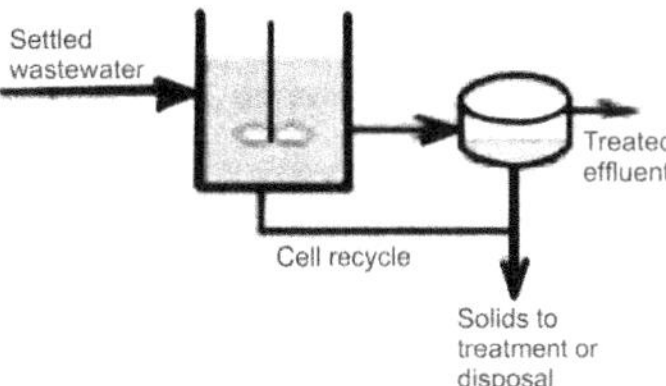

Figure 5.19: Diagram of a Simple Activated Sludge System.

For larger systems, especially when high variability is expected, the design involves the use of multiple aeration tanks and multiple settling tanks. The number of units employed depends on the flow of wastewater being generated.The organic load (generally coming from primary treatment operations such as settling, screening or flotation) enters the reactor where the active microbial population (activated sludge) is present. The reactor must be continuously aerated. The mixture then passes to a secondary settling tank where the cells are settled. The treated wastewater is generally discharged after disinfection while the settled biomass is recycled in part to the aeration basin. The cells must be recycled in order to maintain sufficient biomass to degrade the organic load as quickly as possible. The amount that is recirculated depends on the need to obtain a high degradation rate and on the need for the bacteria to flocculate properly so that the secondary settling separates the cells satisfactorily. As the cells are retained longer in the system, the flocculating characteristics of the cells improve since they start to produce extra cellular slime which favours flocculating.

The most common types of activated sludge are the conventional and the continuous flow stiffed tank (Figure 5.20), in which the contents are completely mixed. In the conventional process, the wastewater is circulated along the aeration tank, with the flow being arranged by baffles in plug flow mode. The oxygen demand for this arrangement is maximum at the inlet as is the organic load concentration.

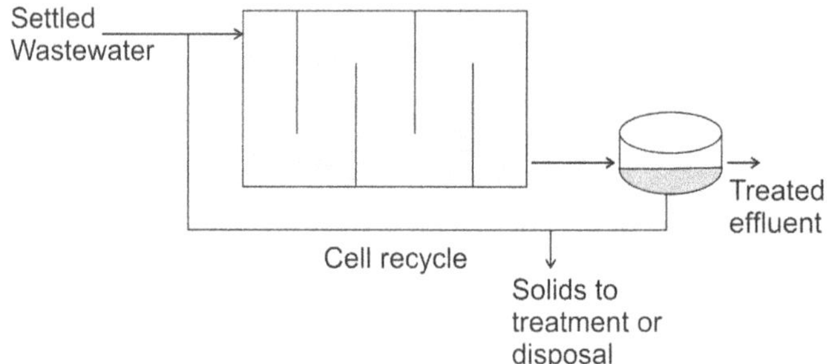

Figure 5.20: Diagram of a Conventional Activated Sludge Process.

In the completely mixed process the inflow streams are usually introduced at several points to facilitate the homogeneity of the mixing; if the mixing is complete, the properties are constant throughout the reactor. This configuration is inherently more stable to perturbations because mixing causes the dilution of the incoming stream into the tank. In fisheries wastewater the perturbations that may appear are peaks of concentration of organic load or flow peaks. The flow peaks can be damped in the primary treatment tanks. The conventional configurations would require less reactor volume if smooth plug flow could be assured, which usually does not occur.

In activated sludge systems, organic load removals of 85-95 per cent are the most common. A key factor in the success of these systems is its proper operation, which requires trained manpower. Although used by some large fisheries which operate on a year-round basis, activated sludge may not prove to be economical or feasible for small seafood processors who operate seasonally because of the need to have a fairly constan supply of wastewater to maintain the micro-organisms.

Problems may appear during the operation of activated sludge systems, including:

- High solids content in clarified effluent, which may be due to too high or too low solids retention time and to growth of filamentous micro-organisms.
- Rising sludge, occurring when sludge that normally settles rises back to the surface after having settled. In most cases, this is caused by the denitrification process, where nitrate present in the effluent is reduced

to nitrogen gas, which then becomes trapped in the sludge causing this to float. This problem can be reduced by decreasing the flow from the aeration basin to the settling tank or reducing the sludge resident time in the settler, either by increasing the rate of recycle to the aeration. basin, increasing the rate of sludge collection from the bottom, or increasing the sludge wasting rate from the system.

- Bulking sludge, that which settles too slowly and is not compactable, caused by the predominance of filamentous organisms. This problem can be due to several factors of which the most common are nutrient balance, wide fluctuations in organic load, oxygen limitation (too low levels), and an improper sludge recycle rate.
- Insufficient reduction of organic load, probably caused by a low solids retention time, insufficient amount of nutrients such as P or N (rare in fisheries wastewaters), short-circuiting in the settling tank, poor mixing in the reactor and insufficient aeration or presence of toxic substances.
- Odours, caused by anaerobic conditions in the settling tanks or insufficient aeration in the reactor.

ii. Aerated Lagoons

The aerated lagoons are basins, normally excavated in earth and operated without solids recycling into the system. This is the major difference with respect to activated sludge systems. Two types are the most common: the completely mixed lagoon (also called completely suspended) in which the concentration of solids and dissolved oxygen are maintained fairly uniform and neither the incoming solids nor the biomass of microorganisms settle, and the facultative (aerobic-anaerobic or partially suspended) lagoons. In the facultative lagoons, the power input is reduced causing accumulation of solids in the bottom which undergo anaerobic decomposition, while the upper portions are maintained aerobic (Figure 5.21 gives an example). The main operational difference between these lagoons is the power input, which is in the order of 2.5-6 Watts per cubic metre (W/m^3) for aerobic lagoons while the requirements for facultative lagoons are of 0.8-1 W/m^3. Being open to the atmosphere, the lagoons are exposed to low temperatures which can cause reduced biological activity and eventually the formation of ice. This can be partially alleviated by increasing the depth of the basin. These units require a secondary sedimentation unit, which in some cases can be a shallow basin excavated in earth, or conventional settling tanks can be used.

If excavated basins are used for settling, care should be taken to provide a residence time long enough for the solids to settle, and there should also be provision for the accumulation of sludge. There is a very high possibility of offensive odour development due to the decomposition of the settled sludge, and algae might develop in the upper layers contributing to an increased content of suspended solids in the effluent. Odours can be minimized by using minimum depths of up to 2 m, while algae production is reduced with liquid retention time of less than two days.

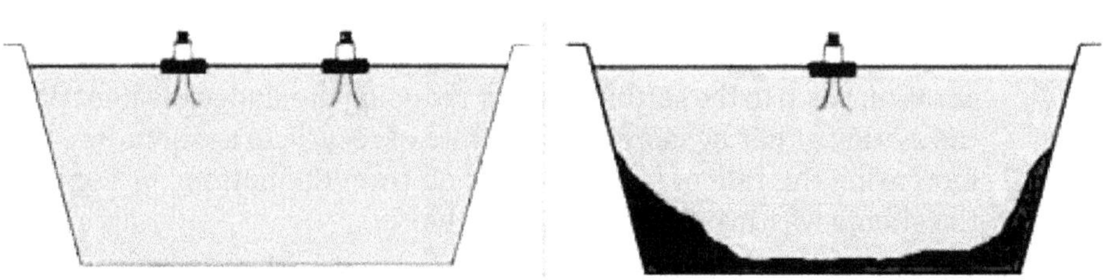

Figure 5.21: Diagram of Aerobic (Top) and Facultative (Bottom) Aerated Lagoons.

The solids will also accumulate, all along the aeration basins in the facultative lagoons and even in comers, or between aeration units in the completely mixed lagoon. These accumulated solids will, on the whole, decomposc in thc bottom, but since there is always a non-biodegradable fraction, a permanent deposit will build up. Therefore, periodic removal of these accumulated solids becomes necessary.

iii. Aeration

The aerated systems described above need an oxygen supply. Depending on the characteristics of the process, different designs may be used. The oxygen can be supplied to the activated sludge by either diffused aeration, by turbine agitation, by static aerators, or by surface coarse or large bubble diffusers. The last two are used also in the lagoon systems.

The diffused aeration systems (Figure 5.22) are also divided into fine bubble, medium and coarse or large bubble diffusers. The fine bubble diffusers are built of porous materials (grains of pure silica or aluminum oxide are bonded ceramically or by resins) which provide very small bubbles of high surface area that favour the oxygen transfer from the air to the wastewater. The medium bubble diffusers

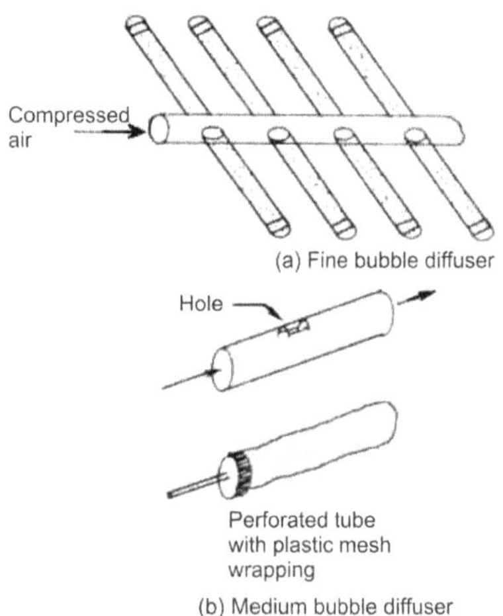

Figure 5.22: Fine and Medium Bubble Diffusers.

are perforated pipes or tubes wrapped with plastic or woven fabric. The coarse or large bubble diffusers can be orifice devices of various types, some of which are designed to be non-clogging (Figure 5.23).

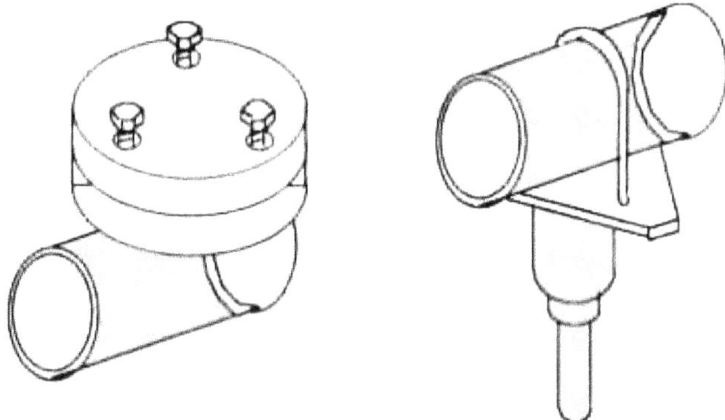

Figure 5.23: Coarse or Large Bubble Diffusers

With the small or fine bubble diffusers, it is important to use air free of particles that would otherwise clog them. Although somewhat less efficient for oxygen transfer, the coarse bubble diffusers are sometimes preferred because the presence of particles in the air is not a critical problem, and also for their lower cost and maintenance requirements. The diffusers are placed along air manifolds, close to the bottom of the aeration tanks.

The static aerators (Figure 5.24) are vertical tubes placed at the bottom of the aeration tank, with packing material along its length. The compressed air is supplied from the bottom of the tubes, forcing a mixture of air and water through the packing, where most of the oxygen transfer to the wastewater takes place. They have been used mainly in aerated lagoons.

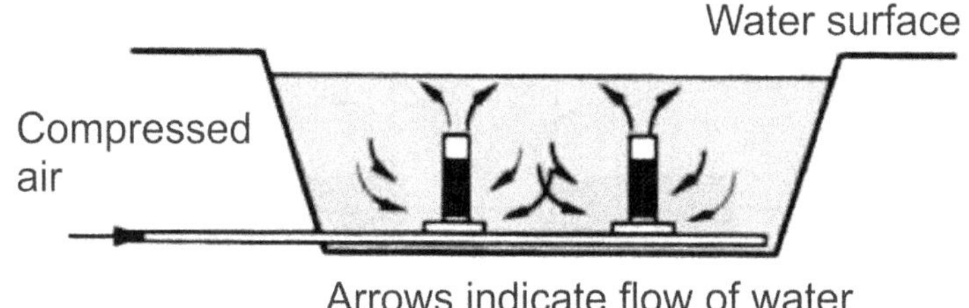

Figure 5.24: Sketch of a Static Aeration System.

The turbine aerators are one of the most common and simple aeration devices and consist of an electric motor-driven turbine impeller rotating at high speed above a pipe or a sparging ring which discharges the compressed air (Figure 5.25). The air bubbles discharged from the pipes are dispersed by the rotation of the turbine. Depending on the depth of the aeration basin, more than one impeller may be used in the same axis. The power drawn by the turbine systems is used for maintaining the micro- organisms in suspension and to break down and disperse the air bubbles, the latter demanding most of the power.

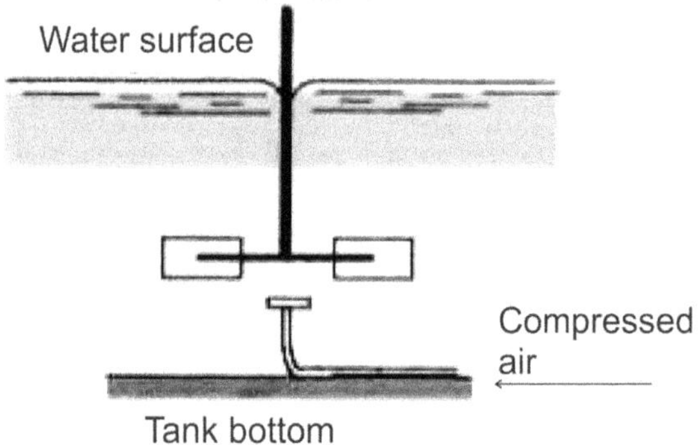

Figure 5.25: Turbine Aeration System.

The most common surface aeration units (Figure 5.26) are mounted on a float and consist of a propeller installed inside a rising tube and driven by a non-immersed motor. The propeller draws the liquid from under the unit and sprays it above the surface of the tank. The oxygen transfer takes place from the air to the droplets sprayed and to the turbulent surface of the liquid surrounding these units. Other surface aeration units are the so-called "brush" aerators which are basically blades mounted on a cylinder which rotates through the liquid (Figure 5.26). Usually, these units require baffles to direct the flow and insure turbulent velocity.

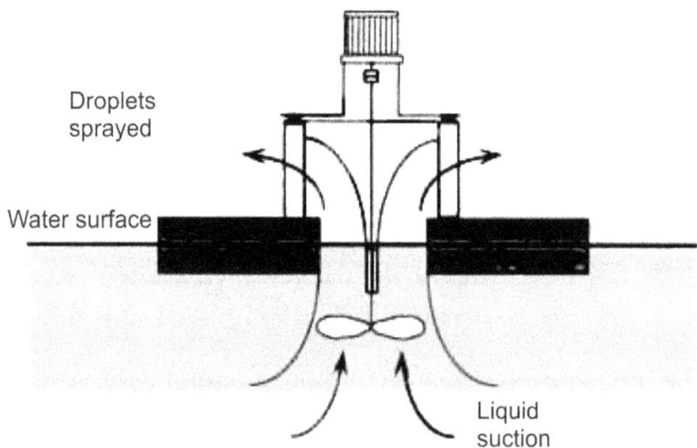

Figure 5.26: Diagram of a Floating Surface Aerator.

Figure 5.27: Sketch of a Surface Brush Aerator.

The oxygen transfer rate of the different devices fluctuate between 0.7 and 1.4 kg of oxygen per kiloWatt-hour when used in actual wastewaters. Most catalogues give much higher transfer capacities, because these values are based on test under standard conditions (typically clean, tap water at 20°C and no dissolved oxygen at the start of the test). When selecting aeration equipment, care should be taken in interpreting these values and transfer rates in actual wastewater should be requested for proper evaluation.

iv. Trickling Filters

The trickling filter is one of the most common attached growth processes. Rather than being suspended as in activated sludge or aerated lagoons, most of the biomass is attached to some support media over which they grow (Figure 5.28).

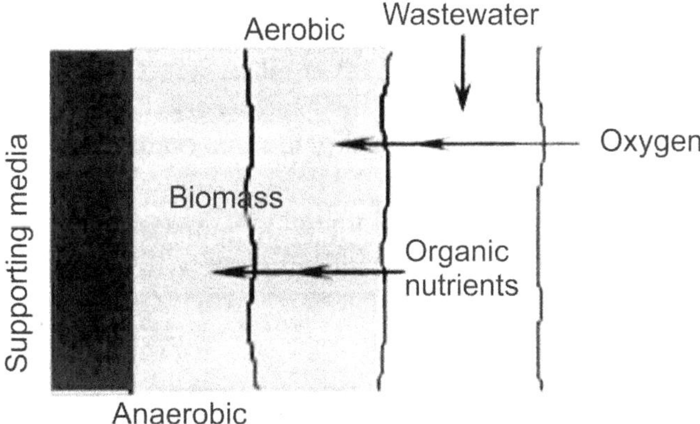

Figure 5.28: Cross-section of an Attached Growth Biomass Film.

The organic contents of the effluents are degraded by the attached growth population which absorbs these organic contents from the surrounding water film. Oxygen from the air diffuses through this liquid film and enters the biomass. As this organic matter grows, the biomass layer becomes thicker and eventually some of the inner portions of thc biomass will be deprived of oxygen or nutrients and will separate from the support media over which a new layer will start to grow.

The separation of biomass occurs in relatively large flocs which settle relatively quickly compared with suspended cells. Air circulates between the interstitial spaces of the supporting material. The media that can be used are beds of rocks (ranging in size from 5 to 10 cm) randomly packed, although regular packings of plastic material (Figure 5.29) are becoming more common recently in view of its much lighter weight, better flow distribution, larger void space and specific area.

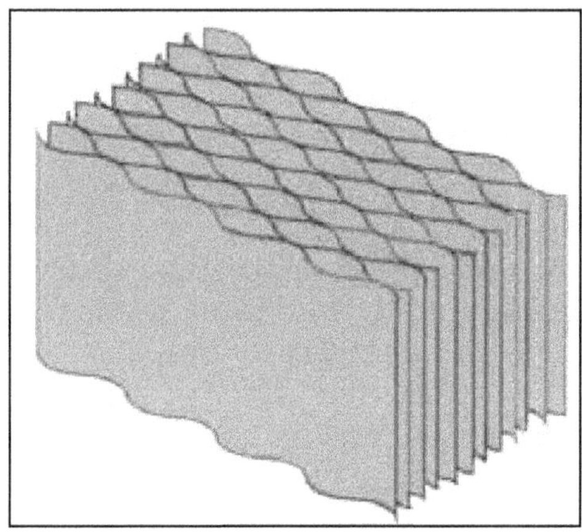

Figure 5.29: Typical Packing for Trickling Filters.

The trickling filter units consist of a circular tank filled with the packing media in depths from 1 to 2.5 m, or 10 m if synthetic packing is used. The bottom of the tank must be constructed rigid enough to support the packing and also designed to collect the treated wastewater which is either sprayed by regularly-spaced nozzles or (more common) by rotating distribution arms (Figure 5.30). The liquid percolates through the packing and the organic load is absorbed and degraded by the biomass while the liquid drains to the bottom where it is collected.

With regard to the packing over which the biomass grows, the void fraction and the specific surface area are important features; the first is necessary to ensure

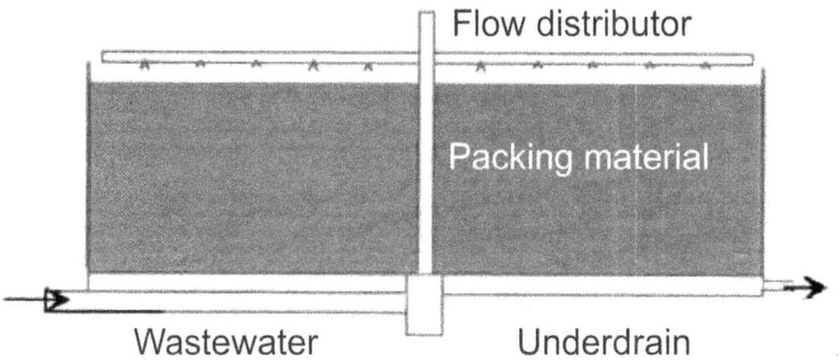

Figure 5.30: Sketch of a Trickling Filter Unit.

a good circulation of air and the second to accommodate as much biomass as possible to degrade the organic load of the wastewaters. Although initially more costly, the synthetic packings have larger void space, larger specific area and are lighter. Usually, the air circulates naturally, but in some high-strength wastewaters forced ventilation is used. They can be used with or without recirculation of the liquid after the settling tank.

The need for recirculation is dictated by the strength of the wastewater and the rate of oxygen transfer to the biomass. Typically, recirculation is used when the BOD_5 of the wastewater to be treated exceeds 500 mg/litre.

As with all biological systems, low temperatures reduce the degrading capacity of trickling filters. In cold areas trickling filters may be covered.The BOD_5 removal efficiency varies with the organic load imposed but usually fluctuates between 45 and 70 per cent for a single-stage filter. Removal efficiencies of up to 90 per cent can be achieved in two stages.

v. Roating Biological Contractors

Rotating biological contractors (RBC) units are another form of attached growth processes. In RBC units the biomass is attached to disks (up to 3.5 m in diameter) which rotate at 1 to 3 rpm while immersed up to 40 per cent in the wastewater (Figure 5.31). The disks are made of corrugated, light plastic material.

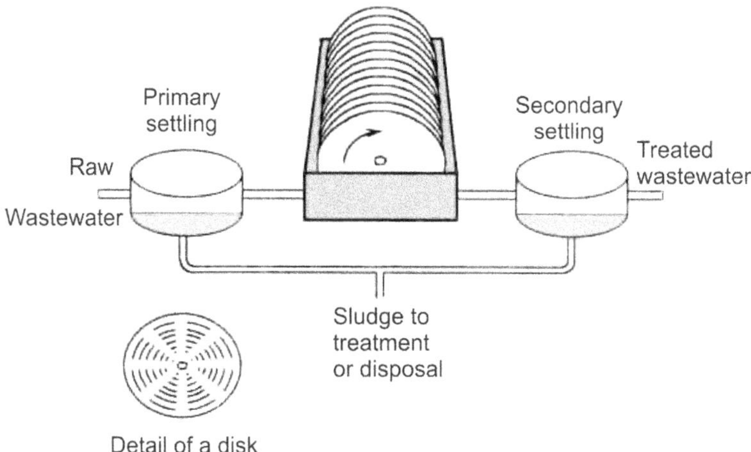

Figure 5.31: Diagram of Rotating Biological Contractor (RBC) Unit.

When exposed to air the attached biomass absorbs air and when immersed the microorganisms absorb the organic load. A biomass of 1-4 mm grows on the surface and its excess is teared off the disks by shearing forces and is separated from the liquid in the secondary settling tank. A small portion of the biomass remains suspended in the liquid within the basin and is also responsible in minor part for the organic load removal. Rotation speeds of more than 3 rpm are seldom used because this increases electric power consumption while the oxygen transfer does not increase sufficiently. The ratio of surface area of disks to liquid volume is typically 5 l/m^2. For high-strength wastewaters, more than one unit in series (staging) is used. The effect of lower temperatures is partially mitigated by the use of housing for the disk units. These systems are normally operated without recycling the liquid. The power consumption is in the order of 2 kW/1 000 m^3/ day of capacity. They have been used to upgrade activated sludge existing plants, placing the disk units in the aeration basins.

vi. Selection of Aerobic Treatments

Several factors (apart from the economics) influence the choice of a particular aerobic treatment system. There is no universal solution and the decision of which system to use (or even if using an aerated system or not) depends on many aspects. Key factors are: the area available, which sometimes is the deciding aspect; the ability to operate intermittently is critical for several fishing industries which do not operate in a continuous fashion or work only seasonally; the skill needed for operation of a particular treatment cannot be neglected; and finally the costs (both operating and initial investment) are also sometimes decisive. The Table 5.4 summarizes these factors when applied to aerobic treatment processes:

Table 5.4: Factors Affecting the Choice of Aerobic Processes

System	*Resistance to Shock Loads of Organics or Toxics*	*Sensitivity to Intermittent Operations*	*Degree of Skill Needed*
	(a) Operating characteristics		
Lagoons	Maximum	Minimum	Minimum
Trickling filters	Moderate	Moderate	Moderate
Activated	Minimum	Maximum	Maximum
System	*Land Needed*	*Initial Costs*	*Operating Costs*
	(b) Cost considerations		
Lagoons	Maximum	Minimum	Minimum
Trickling filters	Moderate	Moderate	Moderate
Activated	Minimum	Maximum	Maximum

The considerations for the RBC systems are similar to those of trickling filters.

5.12.2 Anaerobic Treatment

The anaerobic treatment of wastewater proceeds with degradation of the organic load to gaseous products (mainly methane and carbon dioxide) which constitute most of the reaction products and biomass. Anaerobic treatment is the result of several reactions: the organic load present in the wastewater is first converted to soluble organic material which in turn is consumed by acid producing bacteria to give volatile fatty acids, plus carbon dioxide and hydrogen. The methane producing bacteria consume these to produce methane and carbon dioxide. This is summarized in Figure 5.32.

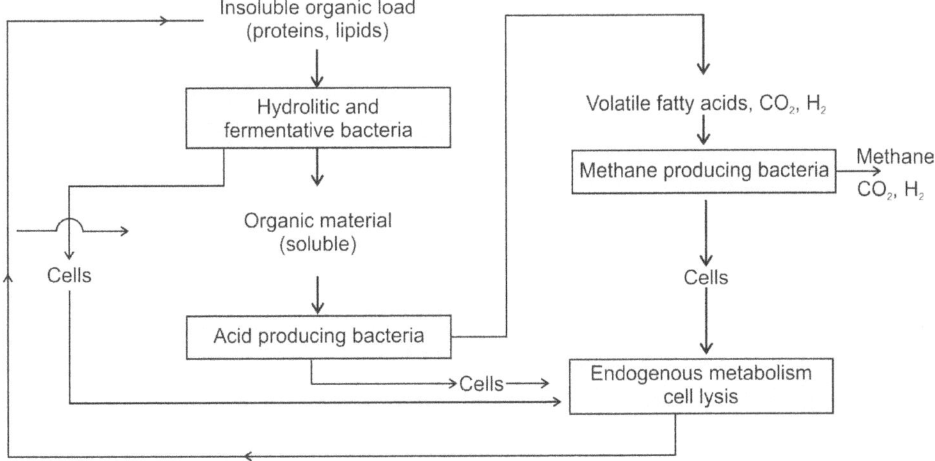

Figure 5.32: Scheme of Reactions Produced during Anaerobic Treatment.

These processes are reported to be better applied to high-strength wastewaters (*e.g.,* blood water or stickwater).

i. Digestion Systems

A typical diagram is shown in Figure 5.33, which presents an anaerobic system.

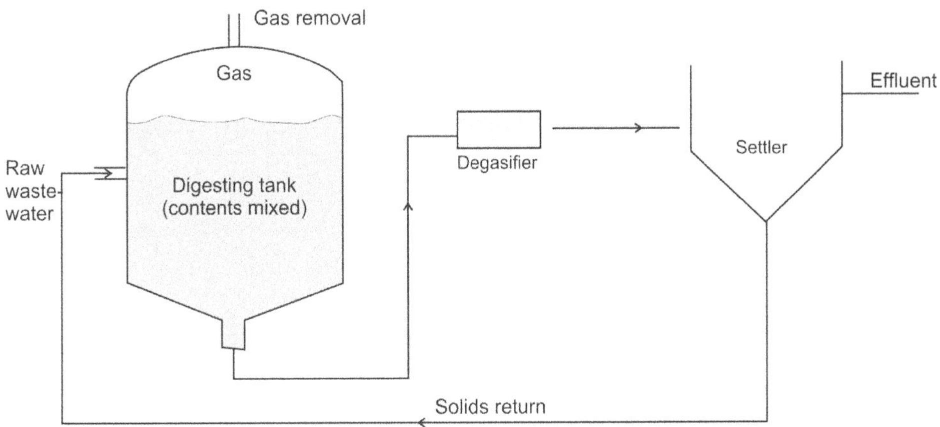

Figure 5.33: Diagram of an Anaerobic Digestion Process.

The flow resembles that of an activated sludge process except that anaerobic digestion occurs due to absence of oxygen. Good sealing of the digesting tanks is essential since oxygen kills some of the anaerobic bacteria present and air presence may easily disrupt the process. From the anaerobic digester the effluent proceeds to a degasifier and to a settler from which the wastewater is discharged and the solids are recycled. The need for recycling appears from the fact that anaerobic digestion proceeds at a much slower rate than aerobic processes, thereby requiring more time and more biomass to achieve high removal efficiencies.

The anaerobic processes have been applied in fisheries wastewaters, obtaining high removal efficiencies (75-80 per cent) with loads of 3 or 4 kg of COD/day/m3 of digester. An alternative process employs a treatment tank filled with packing on which the wastewater is circulated. The bacteria responsible for the anaerobic digestion growth attach on the surface of the packing. The gas produced by a balanced and well-functioning system contains 60-70 per cent of methane, the rest being mostly carbon dioxide and minor amounts of nitrogen and hydrogen. Anaerobic processes are also sensitive to temperature. This is why in some cases heating is provided to the digester to reach temperatures of 30°-35°C. In most cases this can be done in part with the methane gas originating from the digester.

ii. Imhoff Tanks

The Imhoff tank is a relatively simple system used originally instead of heated digesters. It is still used for plants of small capacities. It consists basically of a two-chamber rectangular tank, usually built partially underground (Figure 5.34).

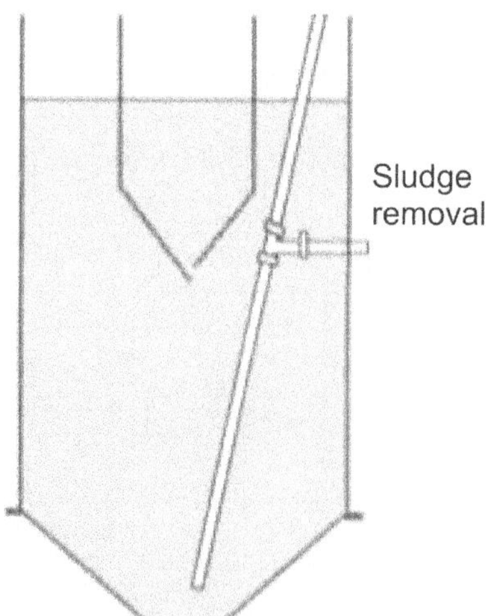

Figure 5.34: Section of an Imhoff Tank.

The wastewater enters in the upper compartment which acts as a settling basin while in the lower part the settled solids are stabilized anaerobically. Short-circuiting of the wastewater can be prevented by using a baffle at the entrance together with more than one port for discharge.

The lower compartment is generally unheated. The stabilized sludge is generally removed from the bottom generally twice a year to give ample time for the sludge to stabilize, although the removal frequency is sometimes dictated by the convenience of sludge disposal. In some cases, these tanks are designed with inlets and outlets at both ends, and the flow of wastewater is reversed periodically so that the sludge in the bottom accumulates evenly. Although they are simple installations, they are not without inconveniences: foaming, odour and scum formation. These typically result when temperature falls below 15°C, when the bacteria that produce volatile acids predominate and methane production is reduced causing a process imbalance. This is why in some cases immersion heaters are used during cold weather. The scum forms because the gases that originate during the anaerobic digestion are entrapped by the solids and have no time to escape from the solids causing them to float. This is usually overcome by increasing the depth in the lower (digestion) chamber. At lower depths, bubbles form at a higher pressure, expand more when rising and are more likely to escape from the solids. The odour problems are minimal when the two stages of the process (acid formation and gas formation) are balanced.

5.13 Physico-chemical Treatments

i. Coagulation - Flocculation

In coagulation operations, a chemical substance is added to an organic colloidal suspension to cause its destabilization by the reduction of forces that keep them apart. It involves the reduction of surface charges responsible for particle repulsions. This reduction in charge causes flocculation (agglomeration). Particles of larger size are then settled and clarified effluent is obtained. A diagram of a coagulation-flocculation and settling of a wastewater is shown in Figure 5.35.

In fisheries wastewaters, the colloids present arc of organic nature and are stabilized by layers of ions that result in particles with the same surface charge, thereby increasing their mutual repulsion and stabilization of the colloidal suspension. This kind of wastewater may contain appreciable amounts of proteins and micro-organisms which become charged due to the ionization of carboxyl and amino groups or their constituent amino acids. The grease and oil particles, which are normally neutral, become charged due to preferential absorption of anions (mainly hydroxyl ions). The parameter that characterizes the stability of a particle is called the Z potential and is a measure of the potential that would be required for destabilization of the particles. However, the usefulness of the Z potential is often questioned since it varies with the composition of the solution and is not repeatable.

The coagulation processes generally employ several steps. First, coagulant is added to the effluent, and mixing proceeds rapidly and with high intensity. The objective is to obtain intimate mixing of the coagulant with the wastewater,

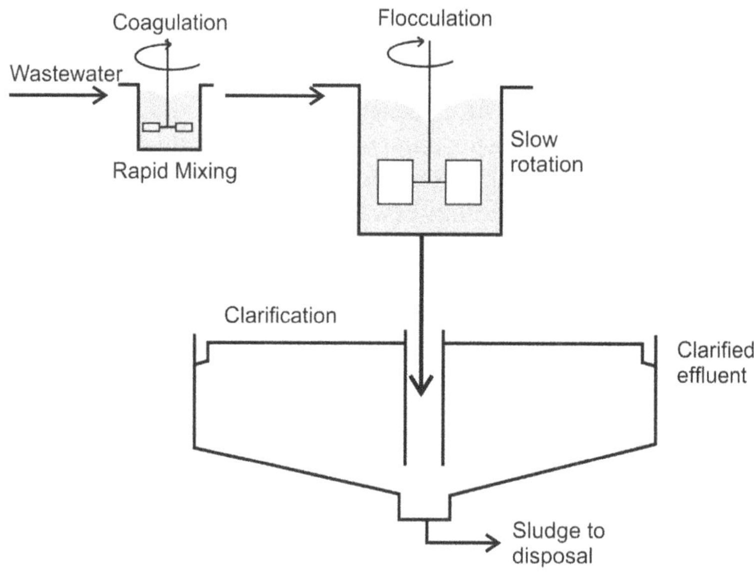

Figure 5.35: Simplified Diagram of a Chemical Coagulation Process.

increasing the effectiveness of destabilization of particles and initiating coagulation. A second stage follows in which flocculation occurs for periods of up to 30 minutes. In the latter case, the suspension is stirred slowly to increase the possibility of contact between coagulating particles and to facilitate the development of large flocs. These flocs are then transferred to a clarification basin in which they settle and are removed from the bottom while the clarified effluent overflows.

Several substances may be used as coagulants. Given the proteinaceous nature of several wastewaters, pH can be adjusted by adding acid or alkali. The first is more common and causes coagulation of the proteins by denaturing them, changing their structural conformation due to the change in charge distribution on their surface. Thermal denaturation of proteins can also be used but due to its high energy demand it is only advisable if excess steam is available. In fact, the "cooking" of the blood- water in fishmeal plants is basically a thermal coagulation process.

Polyelectrolites are also commonly used as coagulants. They act as coagulants by lowering the charge of the wastewater particles. For this case, cationic polyelectrolites are preferred since wastewater particles generally have a negative charge. Some polyelectrolytes act as "bridges" between the already- formed particles. In the flocculation process, bridged particles interact with other bridged particles and this results in the increase of floc size. For this purpose, anionic or neutral polyelectrolites are used. Since the recovered sludges from coagulation-flocculation processes may sometimes be added in the formulation of animal feeds, it is advisable to check that the coagulant or flocculant used are not toxic.

In fisheries wastewaters there are several reports of the use (at both pilot plant and at working scale) of inorganic coagulants such as aluminum sulphate and ferric chloride or ferric sulphate as well as organic coagulants.

There are reports that fish scales can be used effectively as an organic wastewater coagulant. For this purpose, they are dried and ground before being added as coagulant in powder form. Another by-product of marine origin that is used as coagulant is chitosan, a natural polymer derived from chitin, a main constituent of the exoskeletons of crustacea.No single set of operating conditions will satisfy all kinds of fisheries wastewaters and, therefore, the operational policy, pH, and coagulant dosage must be determined for each individual case. This is generally done in the laboratory using beakers and stirrers, trying to simulate the operation at full scale: first, the coagulant is added and rapidly mixed, then stirring proceeds slowly to allow flocculation. Since several variables are studied at a time, it is common to use multiple places (usually 4 or 6) stirrers. These allow constant stirring conditions to be maintained and therefore ascertain the effect of coagulant dosage and pH which is varied in the beakers. The device is commonly known as the "Jar Test apparatus" (Figure 5.36) normally provided with a single speed control and indicator and each stirrer can be either disengaged or placed at different heights.

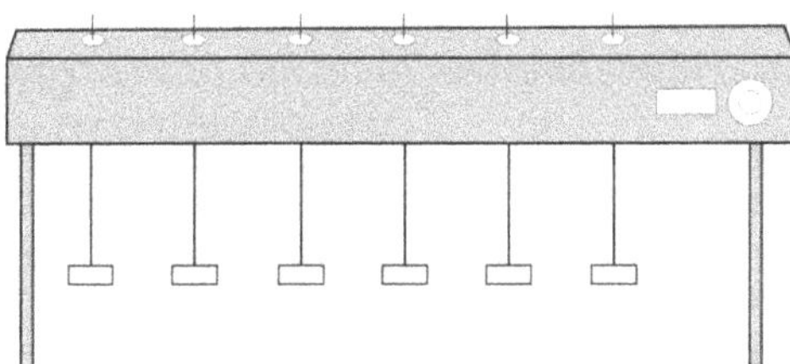

Figure 5.36: Typical Jar Test Apparatus.

5.14 Disinfection

i. Chlorination

Chlorination is a process commonly used in both industrial and domestic wastewaters. Although there may be other reasons for its use in other kinds of wastewaters (*e.g.*, cyanide oxidation), the reason for its use in fisheries effluents is the disinfection by destroying bacteria or algae or inhibiting their growth. Usually the effluents are chlorinated just before their final discharge to the receiving waterbodies.

For this process either chlorine gas or hypochlorite solutions may be used, the latter being easier to handle.

In water solutions, chlorine forms hypochlorous acid which in turn forms hypochlorite: $Cl_2 + H_2O$ « $HOCl + H^+ + Cl^-$ $HOCl$ « $H^+ + OCl^-$

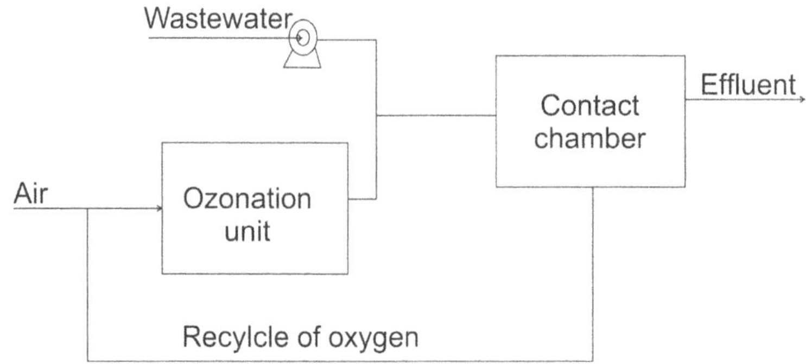

Figure 5.39: Diagram of an Ozonation System.

5.15 Sludge Treatment and Disposal

The sludge generated in the primary and secondary treatments is a suspension of solids in concentrations ranging from about 1 to 10 per cent, depending on the treatment from where it originates. It contains almost all of the organic load removed from the treated wastewater. Although it is in a different form, it is organic matter that may eventually decompose, be offensive and be a contaminant if not treated properly.

In this section, only the most simple treatments will be discussed. ('Me description of more elaborate processes such as chemical conditioning, heat treatment, machine-drying and thermal reduction or incineration can be found elsewhere). Some of the treatments described here may be applicable only by fisheries which have large processing capacities (*e.g.*, digestion), while others (hauling) are more appropriate for smallscale, intermittent operating plants.

5.15.1 Sludge Transportation to Treatment Facilities

The transportation of the sludge generated in plants to sludge treatment facilities is conceptually simple. Since the sludges from fish processing plants are highly biodegradable they will probably be accepted without problems. For small fish processors this solution may only be adopted due to the lack of space for sludge treatment. Other advantages are that no investment is needed for the treatment units and the only operating costs are those of transportation and the fees charged by the remote sludge treatment facility.

5.15.2 Sludge Digestion

Sludges are residues with relatively high carbon content and are good candidates for anaerobic digestion as they are wastes with high organic content. The digesters are very similar to those of anaerobic treatment of wastewaters described previously in section. Here, the cells produced by the anaerobic treatment undergo hydrolysis and enter the process depicted in (Figure 5.40). For sludge digestion, two processes are the most common: the conventional digestion and the high rate digestion, as shown in (Figure 5.40).

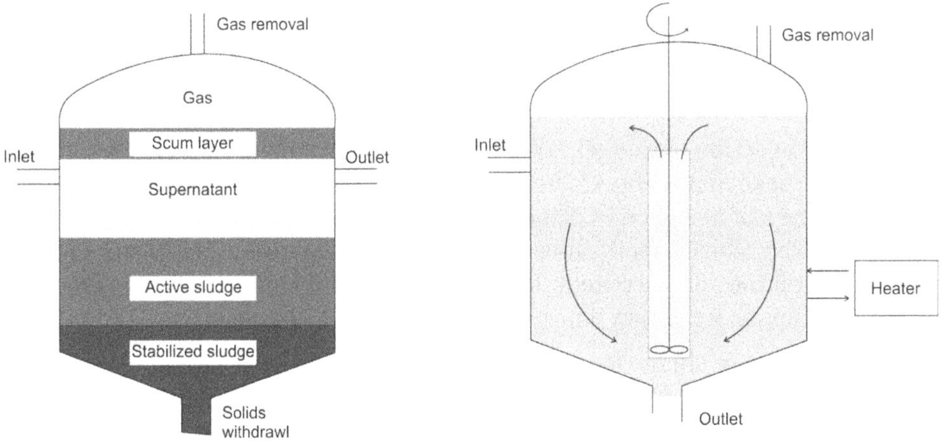

Figure 5.40: Conventional (Left) and High Rate (Right) Sludge Digestion Processes

In the conventional digestion process, sludge digestion and sludge thickening are carried out in the same vessel. The sludge enters the digester in the zone where active digestion is taking place. Due to the conversion of biomass to gases, the sludge stabilizes and settles to the bottom. The gas formed accumulates in the upper part of the digester and, while it rises, carries small particles of oil/grease which form a scum layer on top. Here, the goal is to stabilize the sludge, converting most of it into gas. Typical retention times range from 30 to 60 days and the contents are normally heated to near 35°C in order to accelerate the anaerobic reactions which otherwise would require much more time.

In the high rate sludge digestion, the sludge is mixed by recirculating the gas formed in a draft tube or by pumping. Typical power consumption for mixing ranges from 0.006 to 0.01 kW/m3. Retention times in the digester are shorter (10-20 days) than in the conventional process and higher organic loads can be used. Due to continuous mixing, this process requires a settling tank after the digestion stage.

5.15.3 Land Disposal of Sludges

This is one of the lower cost alternatives for final sludge disposal. It is practised for two reasons: the need to dispose of the sludge (itself also a residue) and the opportunity to use the contents (organic matter, nitrogen and phosphorus) as fertilizer. Sludge land application has proven to be a viable substitute for commercial fertilizer, provided pathogenic control is applied.

Site selection is an important factor; it should be relatively far from waterbodies and the soil should be moderately permeable and well-drained, with a maximum slope of 5-8 per cent in order to minimize erosion problems during application.

Neutral to alkaline soils are preferred to reduce heavy metal mobility, the concentration of which should also be periodically checked.The application of the sludges on the soil can be carried out by spreading from tank trucks or by sprinkling. Trenching or ploughing mixes the soil and sludges adequately, reduces odour and run-off.

5.15.4 Sludge Disinfection

The disinfection of sludges is especially important before they are applied to the land because of possible existence of pathogenic micro-organisms.One of the simplest forms of disinfection is the addition of chlorine to stabilize and disinfect the sludge. The chlorine for disinfection dosage varies between 2 mg/l and 15 mg/l, depending on the solids content. It is usually done in tanks provided with good mixing capability. Retention times range from 15 to 45 minutes.

The storage of sludge for long-term is another low-cost practice, applicable only if enough land is available. The storage time for pathogen reduction depends on temperature: 60 days is recommended if the average temperature is 20°C or 120 days if it is only 4°C.